GROWING UP WITH LUCY

GROWING UP WITH LUCY

How to Build an Android in Twenty Easy Steps

STEVE GRAND

Weidenfeld & Nicolson
LONDON

First published in Great Britain in 2003
by Weidenfeld & Nicolson

A CIP catalogue record for this book
is available from the British Library.

ISBN 0 297 60733 2

Typeset by Selwood Systems,
Midsomer Norton

Printed in Great Britain by Butler & Tanner Ltd,
Frome and London

Weidenfeld & Nicolson

The Orion Publishing Group Ltd
Orion House
5 Upper Saint Martin's Lane
London WC2H 9EA

CONTENTS

I lied about the number of steps

Health warning

There are things in these pages that will give some experts apoplexy. Neuroscientists will be appalled at my over-simplified and caricatured descriptions of the brain and neuroscientific theory. Artificial Intelligence experts will be offended at my offhand dismissal of their heartfelt beliefs. Psychologists will cringe at my simplistic analysis of intelligence. Electronic engineers will scoff at my soldering . . .

I'm sorry about that, but this is a popular science book on a staggeringly complex and varied subject. I can't hope to lay everything out in detail, and to do so would bore most people rigid. The important thing here is the journey through the ideas, and I've tried to give people only enough background to be able to make sense of it. If I have inadvertently offended anybody's sensibilities, then, well, that's tough luck, I guess!

INTRODUCTION

The first time I created life I did it the easy way. At least it seemed pretty easy from my perspective, although for some reason my wife seems to recall it rather differently. This time, though, she can sit back and relax. She doesn't have to put up with morning sickness or back pain, and she won't be required to carry her stomach before her like a galleon in full sail. This time the birth pangs are all mine. This time I'm attempting to make a baby the hard way: one piece at a time.

We living beings are some of the most sublime and complex structures in the known universe, but, as far as I have been able to tell, we are still just mechanisms, no more and no less. If so, then in principle it must be possible to build an artificial being with all of the qualities of a natural one, up to and including self-awareness. This news will not surprise anyone who watches sci-fi movies, and if Hollywood is a reliable prophet we can expect the first android butlers to be delivered any day now. But have you ever really thought about what would need to be done, in order to create a complete living thing from scratch? I don't mean a fancy piece of animatronics or a computer cunningly programmed to behave *as if* it were alive, but something with a real mind of its own that can think and learn and take personal charge of its destiny. Where would we begin? What exactly are the problems and how are they to be solved? Could it ever be done? Should it be allowed? Why would we want to bother?

I expect you have far better things to do with your life than get personally involved in such questions, but I'm sure you'll agree that it's a fascinating notion. In these interdisciplinary times, when biological concepts are infiltrating so many aspects of our technology and culture, it would be interesting to speculate about whether we could really bring together the huge variety of knowledge, skills and ideas that would be needed to make a complete living being. It's not a bad topic for a book, now that I come to think about it.

Of course, idle speculation is one thing but actually to attempt to do it for real would be quite another. After all, we hardly have the faintest idea how the brain works, what the mind does, or what consciousness even *is*; let alone how to replicate any of this. Biological organisms are stupendously complex things, with thousands of millions of working parts, and almost every aspect of the living form is still shot through with mysteries. It must surely be impossible?

Yet how are we going to know if we don't try? And how are we going to prove to ourselves that we understand how natural systems work unless we are able to replicate them? There are hundreds of books on the human brain, extolling the virtues of this or that theory of how it operates, but most of these explanations remain largely conjectural, and those details that can be tested are only ever examined in potentially misleading isolation, as if someone was trying to design a plug without paying much attention to the socket into which it was meant to fit. Many of the particulars are quite well established, to be sure, but to my knowledge there is no over-arching theory of the brain – no description *even of its basic principles of operation* – that can yet be proven to work. And to be frank, this is not terribly surprising. Sometimes it is simply not possible to understand something by taking it apart and examining the pieces, and the human mind is perhaps the epitome of such an irreducible system. If you take an animal apart to see how it works, the first thing you end up with is a heap of bloody entrails that don't work at all any more. Sometimes the only way to understand something is to start with a blank sheet and attempt to build it from scratch, beginning with a pile of bits that 'don't work' and trying to find out how to put them together so that they do. Only then will you know what questions you should really be asking and what problems the system is actually solving. At worst you will come away feeling suitably chastened, with a healthy and informed sense of awe at the beauty and elegance of nature.

So this is precisely what I set out to do. Two years before I started writing this book I laid the foundation stone for an absurdly optimistic project – to build myself a complete artificial creature, whom I christened Lucy, in honour of the famous fossil of one of our hominid ancestors. Lucy is an android robot. Well, to be strictly accurate she's an anthropoid robot, since I modelled her appearance loosely on that of an orang-utan rather than on a human being. In fact, in her first manifestation, Lucy is only half an anthropoid, because she has no movement or sensation below the waist, but she is complete enough to be going on with, and I hope she will pass through a succession of different physical forms over the coming years as my understanding increases, growing up in much the same way that a caterpillar transforms into a butterfly.

Today's technology is still much too primitive to solve all of the problems straightaway, and we currently understand so little about the brain that neuroscience is barely more advanced in theoretical terms than physiology was during the seventeenth century, shortly after the invention of the microscope. Nevertheless, we have to start somewhere, and I think there are good intellectual reasons for trying to tackle the whole of the problem at one go, instead of in isolated pieces and tiny increments. The Lucy project is really a kind of one-man, miniature moonshot – a broad and audacious endeavour with big prizes to be won, but also high personal stakes and a significant chance of complete failure. Yet moonshots are sometimes the only way: you cannot reach the moon by learning to jump incrementally higher and higher, no matter how impressively rapid your initial progress might be; sometimes you need to take a bold step, stop jumping, and set to work on building a moon rocket.

As you will soon discover, I'm not an expert neuroscientist, nor a psychologist, nor a philosopher of mind. I'm just a man-in-the-street: a self-taught tinkerer with a natural desire to understand myself and the world I live in. I'm probably no more qualified to play god than you are, but circumstances seem to have conspired to give me at least a smattering of the skills and experience I need, plus the motivation to try it and an opportunity to have a go. I am also freed from some of the constraints facing professional scientists, who have reputations to maintain, specialisms to stick to and armies of students to keep occupied. Trying to build a practical artificial life form single-handedly is either brave or foolhardy, depending on your perspective. Trying to do it at such an early stage in our understanding undoubtedly raises more questions than it answers, but I think it is worth doing, if only for this reason.

But boy, does it touch a few nerves! In the two years since I started this project I've been accused of being an unscientific charlatan who is only doing this as a get-rich-quick publicity stunt, a heartless reductionist who wants to demean and demote the human spirit, a misogynist who wants to usurp women's role as the nurturer of new life, a hopelessly naïve fool who doesn't know enough science to realise how hard it is and an irresponsible megalomaniac who is trying to bring about the imminent downfall of society. And there was I, foolishly thinking that this was just an innocent personal enquiry into the Nature of Things, motivated by a natural human curiosity and a genuine desire to be helpful. It just goes to show how wrong one can be about one's own motives!

On the other hand, and much to my relief, the majority of people I've spoken to seem to be as fascinated by this project as I am, and I hope you will enjoy reading about my experiences so far. This book is an early progress report – my first letter home from the trenches. The right moment for writing technical research papers will come later, but in the mean time I want to share with you some of the notes and queries that I've been scribbling in the margins of the blueprints as my thoughts have been developing.

I specifically want to address three questions:

First, what can we learn from biology, and what can biological science learn from this? What inspiration can we draw from nature when attempting to design intelligent systems of our own, and conversely, to what extent are we presently being misled by our intuitive and often unquestioned assumptions about how our bodies, brains and minds work?

Second, what can we actually achieve today? Rather than answer this in a hypothetical way, I plan to draw candidly on my experiences with Lucy. At the time of writing Lucy is still far from finished, but already, scruffy though she may be, she is among the most advanced research robots in existence, and her successes and failures stand as indicators of the present state of the art.

Finally, what might we be able to achieve in the future, and what could this mean for us and for our creations? The limitations of today's technology offer hints about the strange materials and machines that we can expect to see tomorrow. And if tomorrow does bring us artificial living beings with real minds of their own, what impact will this have on society, and how will it affect our moral and ethical stance?

I hope you will interpret the stream of consciousness in this book as being a moderate and balanced (while not merely ambivalent or placatory) view of intelligent life, both natural and artificial: creating androids is not as easy as some people seem to think, despite the tongue-in-cheek subtitle of this book, but neither is it an impossible pipedream. Human beings are machines, whether we like it or not, but this doesn't in any way belittle us or rob us of our humanity; it simply means that we need to revise our understanding of machines. Since long before the tale of Frankenstein's monster, the concept of artificial life has both captivated and repelled us, but parables and horror stories are not good guides to literal fact. Dogma, irrational belief and closed-mindedness may well be easier, more comfortable states to adopt (for scientists as much as the general public), but knowing the truth about life is far deeper, richer and more productive. Despite all the hype, artificial intelligence will not lead us to Utopia but neither will it realise our most dystopian nightmares. Like most truths, the reality lies somewhere between these extremes.

From the safety of my perspective, as someone who doesn't actually have to go through childbirth, creating life the natural way seems an almost effortless process: a single cell divides and differentiates, over and over, until somehow it gives rise to a structure whose complexity and beauty utterly boggle the mind. Bearing this precious cargo from the security of the womb into a world large enough to realise its potential is something I'm pretty glad I don't have to experience, but trying to do the same thing by hand, with rivets and solder, isn't exactly relaxing either. I have never had to think so hard in all my life; I could burst a blood vessel any minute, and my labour pains have barely even started yet! There's more to being the parent of an artificial life form than you might think. I shall leave natural childbirth to mothers; let me tell you instead about some of the surprising variety of things one has to consider when attempting to create a baby the hard way.

PART ONE

Conception (and some misconceptions)

Do you suffer from preconceptions? Of course you do. We all do. There's no shame in it. But questioning the unquestioned is a crucial part of the struggle for understanding, and we need to bring our suppositions out into the open.

I'd love to tell you about my own unquestioned assumptions, but I don't know what they are: hence the name. So, in chapter 1 I shall tell you a little about myself, about how I came to be doing this project and what motivates me. This way, you will be able to figure out my prejudices for yourself. It's as well for you to know where I'm coming from, and it will make me feel better about criticising the assumptions of others.

Sadly, even professional scientists have preconceptions that they fail to question. Or sometimes they have carefully identified and formally described axioms that are just plain wrong. Either way, there is a tendency amongst all of us, scientists included, to turn specific examples into incontestable dogma without a moment's thought. People are wont to ask us which school of thought we adhere to, as if there were only a small number of possible options, and acceptance of one tenet of someone's belief implies acceptance of all of them. You say that all tigers are yellow, but I have seen a white tiger; therefore you must be wrong, and all tigers are white . . .

When it comes to artificial intelligence, such misconceptions and polarisations abound amongst experts and laymen alike. Before we can make much progress, we need to examine some of these assumptions.

CHAPTER ONE

Welcome to my mid-life crisis

Why me? Why this? Why now? What kind of an idiot sets out on such a voyage? Perhaps I'd better begin by explaining precisely what kind of an idiot I am.

Picture the scene. A rusting bus stop leans drunkenly against a grey limestone wall, which borders the foot of a sycamore-shaded hill. Strung out to one side of it, twenty pairs of bony knees fidget impatiently in the sunshine, waiting for the bus that will carry them home from school. Half these pairs of knees are complemented by short summer dresses and pretty faces, while the remainder lie almost hidden under drab grey shorts and grubby white shirt tails. Among this group of rather less pretty faces, and thrown into sharp relief against all the monochrome that surrounds it, is a shock of ginger curls, framing a pair of absurd 'Joe-90' spectacles. This is me.

My moment of triumph was about to arrive. For some minutes I had been trying to persuade the two friends standing either side of me that I was not, in fact, an ordinary human being from rural Somerset, but actually an alien envoy from the Moon. To any science-savvy ten year old in the space age of 1968, such a tale would seem perfectly plausible, and my friends were now just about convinced. As a final flourish I removed a small glass bottle from my pocket and brandished it in front of their credulous faces. In the morning this bottle had contained a beautiful homemade 'chemical garden' – delicate towers of gold and blue crystals set in a sea of pearlescent waterglass. Sadly, after a bumpy bus journey to school and a few riotous hours in the playground, all that remained was a foggy brown goo, studded with broken crystalline shards. But it was enough. Only an alien being could possibly own something so magical. Given evidence this conclusive, I almost started to believe the story myself.

One year later, much to my dismay, Armstrong and Aldrin would step out on the lunar surface to be greeted by nothing but a barren, lifeless waste. No sign of a lunar envoy anywhere. Not long afterwards, the glorious Apollo programme would become yesterday's news and the Space Age would slide silently into the history books, carrying many of my fondest childhood hopes with it. In the early days, as I avidly pored over the pictures of American spacecraft in my children's encyclopaedia, I would have bet my life that one day soon I would go to the Moon. Strangely, I was equally certain that I'd never, ever go to America. Thirty years later, as I stood at Cape Canaveral waiting for a rocket to launch, I realised that it was going to go without me, and I'm afraid I cried.

But this disappointment and all the others were still to come. Meanwhile, back in the heady summer of 1968, I could barely wait for the future to start. It was clearly going to be a tremendously exciting time. I was especially looking forward to the arrival of the robots and the Martians. Or maybe they would be Martian robots. Either way, I'd learned so much about them from watching *Fireball XL-5* and *Lost in Space* on television that I was aching to find out what they were made of and how they worked. The comic books of the time sometimes had tantalising cut-away drawings of hypothetical robots, in which parts of their innards were visible for me to pore over and wonder at. And of course, whenever a robot or an alien being was hacked to death by Dan Dare (the justification for which was never questioned in those gung-ho times), there would always be wires and tubes and other grisly bits sticking out of the severed limbs for me to examine. What did all these bits do, I wondered? How did all this complicated clockwork fit together to make something that lived and breathed, and screamed when it died?

Sadly, I was never going to have the chance to find out. Or so I thought.

The thing is, you see, the future kept receding. Back in 1968 when I watched science fiction shows on television, I think I assumed that all this excitement with aliens and androids was actually happening, somewhere across the Atlantic (which was why all the characters had American accents), and it would get to the UK shortly. But gradually it dawned on me that these events were set in another time, known as The Future, which apparently was not due to start for a considerable while yet. Not until the year 2000, according to most pundits. This date was so unimaginably and depressingly distant that I calculated I would be forty-two years old and therefore well into my dotage by the time all the excitement began. Meanwhile, all I could do was wait and dream. And prepare.

And I think I did prepare, too, even though I didn't realise it at the time. Looking back, my life seems intentionally to have been leading up to this moment, although I suppose I should be no more surprised at this than I am to find that my legs are precisely the right length to reach from my hips to the ground. Hindsight is like that.

I started with hopeless little robots of my own invention, built out of Lego or Meccano[1], or sometimes from pieces of old clockwork. They weren't up to the standard of Robbie the Robot on television, it's true, but all that playing with construction toys certainly taught me something important about the nature of building blocks and how, if you get the design of them right, almost everything can be made from almost nothing. This discovery has stayed with me and motivated me ever since. Around the age of eleven, my father introduced me to electronics, beginning with a homemade Christmas present stuffed with interesting experiments involving electricity and magnetism. Again I marvelled at how so much complex behaviour could arise from combinations of so few basic building blocks. On a really good day, this behaviour sometimes even resembled the result I was intending.

Those were the sunny days; the carefree times. Then secondary school came along, and with it the sinking realisation that education simply hadn't been designed with me in mind. There were so many facts for my lazy and erratic brain to memorise, yet so few over-arching principles to tie them together. But I was lucky. Although my spirits sank ever deeper as I failed in the classroom, they soared again during every lunch break and after school. But not for the usual reasons. Whether it was sympathy, indifference or pedagogical astuteness on their part I shall never know; maybe they were just wantonly suicidal, but for some reason I was fortunate enough to have a series of teachers who were willing to let me have more or less free run of the science lab equipment stores out of school hours. Sometimes they even went to the trouble of locking me in! It was in this cloistered and privileged environment that I was able to re-embark on my quest for the ultimate

[1]LEGO® is a registered trademark of the LEGO group
Meccano® is a registered trademark of Meccano SA

building block, beginning at once with the physics lab's radioactive materials store.

A couple of fascinating, glow-in-the-dark, life-shortening years of playing around with protons, neutrons and electrons in this way led me naturally to chemistry, and to nature's ninety-two part construction kit. But it was biology that really made an impact. In the classroom, biology was an endless and bewildering array of pointless names for things. I soon lost track of which part of a leaf was the endoplasmic reticulum and where exactly in the heart one would find the Organ of Corti. But in the lunch break everything was different. The aliens of my childhood turned out to have been amongst us all the time, because the natural world was so much stranger and more bizarre than any thrill-seeking teenager had a right to expect. What's more, all this variety seemed to be constructed from a simple to remember set of building blocks – a few cell types, twenty amino acids and a handful of rules about how genes construct loops of cause and effect that turn simplicity into complexity. And somewhere, dimly perceived, was an even more profound set of building blocks that nobody quite seemed to understand.

This was when I made my first major blunder. I realise now that by the age of seventeen I had started to become interested in the physical building blocks of the mind. I was intrigued by the mystery of how children's minds manage to emerge all by themselves out of a squishy, pinkish substance with the approximate consistency of mozzarella. But I couldn't quite locate this feeling, and somehow I managed to convince myself that I wanted to be a schoolteacher. This was a big mistake. I was shy, and often felt self-conscious just going into a shop to buy sweets, so having to act out children's stories in front of thirty ruthless nine year olds and a succession of stony-faced supervisors ('just pretend I'm not here') was enough to give me the heebie-jeebies. By the time I realised my error and owned up to it, it was too late to switch courses and get a degree. Thus ended my higher education.

Happily, serendipity once again stepped in and saved the day, this time by introducing me to a newly invented toy called the micro-computer. It was painfully obvious that I couldn't teach children, but it turned out that I was pretty good at teaching computers. I even found that I could teach computers how to learn things for themselves. And so one thing led inexorably to another over the course of the next decade until, to cut a very long story short, I woke up one day in the late nineteen-eighties to discover that I had become a computer games programmer.

I remember sitting on a train once, listening to some kid explaining to a complete stranger how he planned to accomplish his life's ambition to 'break into the computer games business'. In my case I broke into it like an overweight skater into a frozen pond: it was an unfortunate accident. I was far too old by this time to think that writing video games was a cool thing to do, and quite honestly all that joystick twiddling just wore me out, but after a few false starts I managed to find my niche. I heaped everything I'd learned up to that point about the building blocks of life into a computer game (of sorts) called *Creatures*, which allowed people to keep artificial life forms as pets. Writing it was quite cathartic; a pouring out of scientific ideas and computer experiments that I had kept bottled up for years, for want of somewhere to deploy them.

Much to my amazement the game turned out to be a success, and I found myself elevated from the lowly position of programmer to that of technical director of a new company, aimed at exploiting these ideas for more serious and useful applications. At first it felt wonderful to have some status and influence after all those years of scratching a living and having my ideas ignored. Suddenly everyone from big corporations to the media was excited about the potential for this way of thinking about artificial intelligence, and our prospects looked good. But gradually and relentlessly it all went wrong.

The tale is so familiar that you will probably recognise the cast instantly: people who are good at working with their hands but suddenly find themselves given a suit and told to be managers (just like teaching, only worse); people who understand the bottom line clearly enough, but whose technical illiteracy and ignorance of the creative process mean that everything they do stifles the baby they so want to nurture, and finally those who think that many hands make light work (and bigger empires), when in reality they just create more mouths to feed . . .

And so we parted company. My skills were clearly not valued there any more, and suddenly I found I had become redundant. It wasn't a great surprise, but nevertheless this sort of thing is not good for one's self-esteem, never mind one's bank balance. The dream had suddenly turned sour and my soul was all but destroyed (again). I thought I was going to have to abandon my passion for artificial life completely, and

I would have done so there and then, quite frankly, if it hadn't been for the OBE.

I didn't see that one coming at all. We had just arrived back from holiday and I had a terrible migraine in anticipation of all the corporate angst to come, so I went straight to bed. An hour or two later, once Ann could no longer resist opening the letter marked '10 Downing Street' that had been sitting among the heap on the doormat, she woke me and told me the news. In my befuddled state this amazing information hit me with all the force of a gnat landing on butter. 'Oh. That's nice,' I muttered vacantly from a point in the middle distance, and then turned over and went back to sleep to escape the millions of people with pneumatic drills who were trying to break out of my skull.

It was nice, too. I honestly don't mean to boast – I'm under no illusions about the arbitrariness of the British honours system. I think I simply hit lucky when the things I was trying to achieve just happened to coincide with something the government was keen to promote. I must say I felt rather a fraud going to the Palace to collect my medal, amongst all those people who had devoted their lives to charitable works or made a name for themselves in some soap opera. 'And do you write those horrid little computer games?' the Queen asked. 'Guilty as charged, Ma'am' said I, or so it seems in my recollection. But the timing was perfect. Within the space of a few weeks I had been rejected by my company and rewarded by my country. It all seemed to balance out quite satisfactorily. I know it's only a little pink ribbon with a gold cross dangling from it. I'm sure I shall never wear it – pink isn't my colour. But the point is, in my darkest hour, somebody somewhere thought I was doing something worthwhile, and that meant a lot to me.

And get this: the official moment at which 'Steve Grand, failed teacher and useless businessman' turned seamlessly into 'Steve Grand, Officer of the Most Excellent Order of the (admittedly non-existent) British Empire' was midday on the 31st of December 1999 – the very verge of a new millennium and, as I'd learned so agonisingly as a child, a mere *twelve hours* before The Future was officially due to start. I think it was a sign.

So what choice did I have? We have reached the future already and the robots and Martians have flatly refused to come to Muhammad, so Muhammad will simply have to go to them. I can't let all of those childhood questions remain unanswered. My impetus has always been two-fold, and combines a love of analysis with a respect for synthesis. I want to understand the extremely simple and elegant building blocks that underlie most natural phenomena, but also to celebrate the endless richness that arises when they are put together in the right ways. My particular obsession with understanding the building blocks of the mind has only become stronger over the years, and after decades of pondering and experimenting I feel I may actually be on to something important at last. This is why, at an age when I am supposed to be responsible and respectable and have conversations about investment portfolios, I decided instead to have a mid-life crisis. With the kind indulgence of my wife Ann and my son Christopher, I dropped out for a while, to see if I could fritter away our hard-earned life savings by doing something radical, marvellous and, almost without question, utterly futile.

I began to build myself a daughter. Her name is Lucy.

CHAPTER TWO

The mythical algorithm of thought

This isn't as obvious or well known as it should be, so let me make it clear right from the start: if we knew how to make machines intelligent, we would have done it long ago. Rather than confidently striding forth along a lengthy but well-marked path, it seems to me we are stuck half-way up a dead-end creek without a paddle. Remember what I said about trying to reach the moon by jumping higher and higher? We can jump much higher today than we could when artificial intelligence research began, nearly half a century ago. This sounds encouragingly like progress, but in truth I don't think we will ever get to the moon that way.

Amazingly, though, some still believe they could, if only they had access to better training shoes. Essays about machine intelligence (including this one, it seems) sometimes begin by quoting Moore's Law, which states that computer technology tends to double in power about every eighteen months. From this, the authors extrapolate that in hardly any time at all we will have machines with more computing power than the human brain (whatever that means), at which point, they conclude, we will inevitably find ourselves saddled with hyper-intelligent machines, whose only conceivable interests in life will be to Enslave Mankind and Rule the Earth.

Now, I can certainly accept the validity of Moore's Law – how else can I justify the ridiculous amount of money I've spent on trying to keep up with the latest in PC technology over the years? But to imply that powerful computers lead automatically to more intelligent software is of course just plain nonsense. If I gave these doomsayers a computer that was a billion times faster than those of today, and had a billion times more memory, would it spontaneously begin thinking up plans for world domination the second they switched it on? No, of course not. It would just sit there until it was programmed in the right way. And how are they planning to do that, exactly?

When you think about it, there are really only two plausible answers to this question: either the computer has to be programmed to behave like a brain, or it has to be programmed to behave like a mind. What is the difference, you might ask? Surely the one gives rise to the other? Well, yes. Or at least, some brains give rise to minds. But the mind also has a certain logic of its own, which we might plausibly hope to reproduce without needing to worry about the messy details of how it is implemented in the brain. If this weren't true, at least to some degree, then an awful lot of psychologists would suddenly find themselves out of a job, because psychology is the study of how the mind operates, and it more or less goes without saying that unless the mind operates in something like a lawful and predictable manner, this would be impossible.

In fact, some would say that 'a certain logic of its own' is an especially apt phrase. We sometimes appear to be thinking in a logical way about things, one step after another, as if obeying the sort of Rules for Reasoning that the Ancient Greeks used to enjoy dreaming up during dinner parties. I say we *appear* to do these things, because the evidence suggests that we do nothing of the kind – we simply aren't as rational as we like to think. Nevertheless, the philosophers do seem to have the rules for logical thinking pretty much nailed down now, and during the nineteenth century, the mathematician George Boole even invented a complete algebra for automating logical thought, so perhaps we could program these rules straight into a computer (which is after all a logical device, based directly on Boolean algebra) and avoid all that messing around with neurons. Why not simply cut out the middleman?

Another way of looking at this is to define the mind as a 'virtual machine' that is created by the brain, in much the same sense that a word processor is a 'virtual typewriter' created by a computer. If computers can emulate other kinds of machine – and when you think about it, many software applications are analogues of other physical machines, such as filing cabinets, calculators and aircraft (in flight simulators) – then perhaps a computer can emulate the mind just as well as the brain can.

This is certainly close to the way of thinking about the problem that workers in artificial intelligence have used most often. It relies on an idea called computational functionalism, which asserts that if two systems with different structures behave in the same way, then they are truly equivalent, despite their internal differences. So, the argument goes, if we can program the logic of the mind straight into a computer, the machine will really think, even though a computer program looks nothing like a brain.

In itself, this is hard to deny. Suppose I show you two large boxes, each furnished with a microphone and loudspeaker. I tell you that one box contains a human being and the other a devilishly clever piece of clockwork. You ask questions into each microphone in turn and find, much to your surprise, that both boxes respond with equally sensible answers. If this continues indefinitely then you will be unable to tell me which box contains the person and which is the mechanical marvel. And it really doesn't matter. Not only is the quality of thought equal in both cases, even though one set of answers emanates from a glorified alarm clock, but also, unless you can open the boxes and look inside them, there is no sense in which a difference even *exists*, as far as you are concerned. The internal workings remain as inscrutable as Schrödinger's famous cat. Since the internal differences can have no external consequences, they are irrelevant, and we must conclude that the clockwork machine is really thinking. This, of course, is the essence of the famous Turing test for artificial intelligence.

But (and you knew there would be a 'but', didn't you?) this only applies if the behaviour of the two systems is truly comparable over a broad range of circumstances. In the case of artificial intelligence some differences between the outputs of the natural and artificial systems would obviously be acceptable, in exactly the same way that a chimpanzee and a gorilla show differences in their responses, but not enough to make us suspect that one or the other is only *pretending* to think. Nevertheless, if the difference grew too large or lacked certain crucial characteristics, we would be right in concluding that the disparity was caused by something unacceptably askew inside the artificial system. In the worst case we would say that the artificial system is to the natural one no more than a portrait is to a person – a shallow imitation of the real thing.

In practice the snag is that nobody has yet managed to formulate a description of the rules for thought that is complete and generally applicable. So far the various approaches have worked in one very narrow domain, but then failed miserably in all the others. This is because, although there is nothing intrinsically wrong with the functionalist principle of equivalence, there may be something wrong with the assumption that there are many ways of making a mind.

The difficulty with relying upon functionalism in AI is that abandoning the brain leaves us nothing to draw our inspiration from. Given all the different kinds of program we could possibly write, how do we know which of them will be intelligent? We might assume, for example, that thinking is the process of applying logical rules, but sometimes thought isn't logical. And anyway, how can logic be used to control all the other intelligent but unthinking things that brains do, like learning to ride a bicycle? – 'I seem to be swerving towards that car. Let me see now: if my front wheel is in state *x* and my balance is in state *y*, then that means I should ... oh shit!'

Our early geocentric theories of the solar system faced a similar problem. Astronomers would keep uncovering awkward new facts, starting with the retrograde motion of the planets, so the theory constantly needed to have special twiddly bits added to it, in an impressively baroque manner. Eventually it all broke down under the strain and it became clear that it had been fundamentally flawed right from the start. It wasn't that the early astronomers were stupid; just that it hadn't really occurred to them that placing the Earth at the centre of the heavens was an *assumption*, and one that might not be correct. In much the same way, trying to build an intelligent system without reference to the one-and-only class of machine we know of that generates intelligence in nature quickly leads to the bolting on of endless incompatible solutions to deal with special cases, suggesting that the underlying assumptions may not be true. It may be that the logic of the mind can be fully described without reference to the logic of the brain, but at the moment we don't know how. Everybody's solution is different, and none of them is anywhere near general enough to do everything an intelligent creature can do. After fifty years of relatively unproductive effort, only one thing seems certain: the mind does not, in any meaningful sense, work like a computer program.

Which is a bit unfortunate, really, since the digital computer was largely intended to be an analogue of the mind. We don't talk about computer 'memory' and 'instructions' and 'logic' for nothing: these metaphors are drawn from the way that psychologists had already chosen to divide up the structure of the human mind. We appear to memorise and recall things, and so it was assumed that there is a place called a 'memory' in our brains, in which these things may be stored and retrieved. The things that we store and retrieve there can be described collectively as knowledge, or specifically as facts and rules, which came circuitously into the jargon of computing via mathematics (another abstraction of the process of thought), and hence are usually referred to as 'data' and 'functions'. If the process we call thought involves the acquisition and utilisation of knowledge, then it was assumed that we must have a location in which these rules and facts, having been retrieved from the place called memory, are combined to carry out the process we call thinking. In a computer this location is called the 'central processing unit'.

Alas, just because something complex *can* be broken down into functionally distinct units in a certain way, this alone is no reason to suppose it is *actually* divided up in that way. And just because something has a specific name, we shouldn't assume it also has a specific location. This failure to distinguish divisions that actually exist in the world from those that are simply convenient inventions to help us think about things is a common and dangerous weakness of the human intellect. We fail to distinguish between innate and imposed categories because classifying and pigeonholing the world is something that our brains are powerfully driven to do. Perhaps I should conclude from this that we have a special 'classification module' somewhere inside our heads!

Just as the architecture of the computer was divided up into memory, processor, program and data, on the dubious authority of prevailing theories of mind, so too has AI generally divided up intelligence into apparently logical units such as vision, knowledge management, planners, speech recognisers, natural language parsers and so on. And then the researchers have set off and tried to emulate each of these supposed structures or processes individually, each time using a different method, on the implicit assumption that they can be bolted together again later to produce something with general intelligence. Alternatively, now that this anticipated synthesis has turned out to be much more difficult than anyone expected, the modules are allowed to remain separate, with the explanation that each one constitutes an intelligent system in its own right. This is partly how the marketing guys manage to get away with claiming that microwave ovens, speech-dial mobile phones and optical character readers contain 'artificial intelligence'.

Actually, this usage of the term brings to mind another widely held misconception about AI. Artificial intelligence is frequently defined as 'enabling machines to perform tasks which, when performed by a human being, require intelligence'. Now, this is a worthy aim and I'm not knocking it. I have to use my intelligence to add up a shopping bill in my head, so I'm very grateful that I have a pocket calculator to do it for me. I require intelligence to check my spellings, so I'm thankful for the invention of the spell-checker. I'm even glad of the existence of machines that can play chess, which therefore save me the considerable trouble of having to play it myself. But beware. I also require intelligence to stand on one leg – I know this because if I stop thinking about it or make an error of judgement I fall over. A telegraph pole can stand on one leg for much longer than I can, but this is not normally held as evidence that telegraph poles are smarter than me. Making machines that can do what we use intelligence to do is therefore not the same thing as making machines intelligent.

In fact I find adding up, spelling and board games relatively difficult things to do, precisely *because* I'm intelligent: it is the possession of intelligence that makes me into a general-purpose machine, which can adapt to new circumstances and solve unfamiliar problems. Like any general-purpose device, there will always be a specialised solution that can outperform me on a single task, but this is easily outweighed by my qualities of adaptability, flexibility and robustness. Unfortunately, many of the things that march under the banner of artificial intelligence, worthy as they are, should properly be described as *substitutes* for intelligence, and no matter how useful they are, the term AI is something of a misnomer. What I'm interested in with Lucy is capturing this 'general-purposeness' and it's as well to be aware that by no means all AI research shares this aim.

But to return to the business of dividing up the mind into functionally distinct modules: such classification schemes might be reasonable, but then again they might not, and if they are not, then they could be extremely misleading. Consider the parable of the blind men and the elephant. One man feels the elephant's trunk and declares the animal to be like a snake, another feels its leg and claims it to be like a tree, a third feels its ears and concludes that elephants are like fans. Never mind the original point of the story – if the blind men get together and share notes, they will come up with a fairly complete and possibly useful description of the beast. But suppose they now decide to build an elephant for themselves. Taking a python and a pair of large fans and strapping them on to four vertical tree trunks might end up looking a bit like an elephant, but not in any way that another elephant would appreciate!

This is not as flippant as it might seem. A functional analysis of the mind is an abstraction, and abstractions are useful, if not essential, aids to understanding. But abstractions always lose something, and you can no more build a working system from abstractions than you can fight a real war using chessmen. The operation 'is like' is not necessarily commutative: chess is like a battle, but battles are not taken in turns; the heart is like a pump, but bicycle pumps are no substitute for hearts; the computer is like a brain, but the brain need not be divided into separate memory and processor units like a computer. Sometimes one system can genuinely be substituted for another, when both of them share a common abstract description, but not always by any means, and it depends very much on how you make the abstraction. At any rate, no matter how much correspondence there may seem to be between what a computer does and what a brain does, it doesn't follow that intelligence can be 'programmed in'.

Yet still people persist in asking me what Lucy will be 'programmed to do'. Sometimes I simply have to leave the room and scream, because after several years of being asked the same question about my earlier attempts at artificial life I still can't think of anything succinct and witty to say that will clear up the confusion. It seems so obvious to them: first you build a computerised robot, and then you program it to do something. The trouble is that this is an accurate statement about the vast majority of robots that exist, from the huge industrial robots that spray-paint cars, to those that are superficially similar to Lucy, such as Sony's robot pet dog, Aibo. But Lucy is not like this at all, even though her behaviour will ultimately be the result of a set of computer programs. It partly comes down to how you perceive computers, and I think most people perceive them the wrong way. I can feel another flashback coming on ...

The first computer program I ever wrote was to compose poetry. Heaven knows why – I write awful poetry and the program was much, much worse than me. It just seemed like a nice easy exercise to help me learn my first programming language and was rather more interesting than writing a payroll program, which was the usual way that educators tried to put people off computing. When it came to running the program, I came up against a slight difficulty – I didn't own a computer. Instead I had to pretend to be the computer myself, blindly carrying out the instructions I had written, to see what came out (and trust me, you don't want to read what came out!). But by 1978 I had built my first real computer from a kit (a Nascom 1) and after a few small embarrassments such as placing every single chip in its socket the wrong way round, I eagerly set to and programmed the machine to learn how to play draughts (checkers).

The salient point about both these programs is that they are process-oriented. One of them performs the process of writing awful poetry and the other carries out the process of playing draughts. Digital computers are specifically meant to be machines that carry out sequences of actions, and so I took it for granted that sensible applications for computers would also involve sequential processes. Then one day I saw a program that completely changed my understanding of computers. It was written by Willie Crowther and Don Woods, and was called *Adventure*.

> You are standing at the end of a road before a small brick building. Around you is a forest. A small stream flows out of the building and down a gully. In the distance there is a tall gleaming white tower.

These were the enigmatic opening lines of the first ever computer adventure game, which did its best (given the limitations of the time) to simulate a complete underground world, full of magical tricks and traps for the player to overcome. You interacted with the virtual world by typing simple commands, such as 'go north' or 'get lamp' but, beyond this, the things you tried to do and the order in which you did them were more or less up to you. In practice the program was barely more than a branching story, with the added twist of a stored internal state: if you dropped the lamp in a certain location it would still be there when you went back. Nevertheless, I was deeply intrigued by the revelation that a computer could not only execute processes but also represent a place, containing objects. Computers could not only do things, they could *be* things.

I started to wonder whether the objects in an adventure game could be given arbitrary properties, besides a name and a location, in such a way that the objects could sensibly interact with each other without the programmer having to consider every possible combination and its outcome in advance. The behaviour of such a system would of course be the direct consequence of whatever was programmed into it, and yet nowhere in the program would you be able to find an explicit description of that behaviour. It would have *emergent properties* that the programmer neither deliberately specified, nor necessarily expected.

This rather appealed to the anarchist in me, and it was such an enticing idea that I sat down and wrote my own adventure engine, called Microcosm. Over the next few years I used this to create a series of adventures for schoolchildren, allowing them to meet people and explore places and times that would otherwise be out of their reach. The idea was that the children would not be constrained by my interpretation of the story-line but would be able to have adventures entirely of their own making.

Since I'm self-taught I do tend to reinvent the wheel quite a lot, usually in blissful ignorance that other people have passed this way long before me. In this particular case, part of what I had stumbled on was already known as object-oriented programming, which is conceptually quite different from procedural (process-oriented) programming. In itself this wasn't a new discovery (it simply hadn't yet become fashionable) but the important thing – the mindset-changing thing – was that I henceforth stopped thinking of a computer as a device for carrying out instructions and executing a process and started to think of it as a place in which to build things. I have never looked back. The paradigm of algorithms and functions and processes gave way to one comprising objects and properties and structures. I simply can't tell you how much this changed everything in my mind. It took me right back to the days when I used to build toy robots from Meccano, and invariably wished that I had a type of component that didn't exist. Now I found I had a machine where I could build anything I liked – things that were too difficult to make in the physical world, or even things with physically impossible properties. And then I could plug these virtual building blocks together in cyberspace to make larger structures, whose behaviour would be greater than the sum of their parts.

This is an idea that underpins the alternative way of looking at artificial intelligence that I mentioned at the start of this chapter.

Instead of trying to emulate the mind directly with a computer program, perhaps we can emulate the structure of the brain instead, and let the mind emerge all by itself. Rather than use the digital computer as a logical processing device to run some hypothetical 'algorithm of thought', we can use it as a substitute *material*, from which we can construct virtual neurons and maybe even virtual chemicals and other biological building blocks inside cyberspace. Then we can plug these together into large networks that emulate the organisation of some kind of biological brain. This approach to AI is known as Connectionism, and the structures created in this way are called neural networks.

There is, however, just one tiny, niggling problem: we don't know what the detailed structure of the brain is like, and even if we did, the human brain (or any mammal's brain) is a stupendously large network of neurons by anybody's standard. So in practice the connectionists did what any sane person would do in the circumstances. They took one look at the hundred thousand million neurons in the human brain and said to themselves: 'You must be kidding! Maybe we'd better lower our sights a little.' And essentially that's what they did, because if most of today's neural networks are inspired by anything at all (and sometimes it's hard to tell) then it certainly isn't the structure of the human brain.

CHAPTER THREE

Cogs to cockroaches

The last few clattering footsteps sulk away into distant corridors and a hush descends upon the small crowd of upturned faces. A man in a priestly gown glances surreptitiously at his watch and shakes it slightly, as if any discrepancy in timing must be the fault of modern quartz or silicon, and nothing to do with the six hundred years that have passed since this spectacle was first set in motion. For a moment there is nothing more to occupy the tourists' senses than the musty odour of a thousand years of English sweat, incense and death. And then it begins: high on the wall, heralded by a whirr and a clunk, a sullen creature by the name of Jack Blandiver starts to tap out the hour on a bell with his stick. Some metres to the left, a succession of mediaeval knights begin to carousel past each other, one of them being knocked repeatedly from his horse by a lance, apparently having learned nothing from the previous twenty million times this has happened to him. Those in the crowd who can decipher the complex and ornate dial on the wall will see that it is now officially eleven a.m., and the Moon is in its third quarter. For the benefit of any not versed in the art of reading, the cathedral bell also tolls out the time, and then silence returns to my tiny home town of Wells, in Somerset, as one of the world's oldest computational machines goes back to counting silently on its fingers.

Until recently, clockwork like this was almost our only way of carrying out mechanical computations. Simple trains of cogs can easily perform multiplication and division, and with a little ingenuity they can produce a variety of more sophisticated linear transformations, such as the calculation of differences in a differential gearbox. Adding special non-linear elements to the toolkit, such as the ratchet and the escapement, makes far more complex processes possible, for example the self-regulating pendulum clock or the complex calculations (almost) made possible by Charles Babbage's revolutionary analytical engine.

During the twentieth century, we passed from the age of clockwork into the age of analogue electronics, but this was essentially a solid-state manifestation of the same ideas. Radio and television are (or at least used to be) analogue devices, which perform fixed computations on electrical signals using transistors in place of cogs.[1] For a brief and under-appreciated moment we even had analogue computers, in which sets of electronic 'cogs' and 'ratchets' could quickly be rearranged to perform arbitrary mathematical transformations, using electrical voltages to represent numerical quantities.

Then came the digital logic revolution. Digital logic is based on the logical algebra invented by George Boole, but it encapsulates these logical operations in physical form as small transistor circuits called gates. Any question about the truth of something can be determined using an equation in which other truth-values (TRUE or FALSE) are combined by the logical operators AND, OR and NOT. So by substituting the actual truth of each fact into the algebraic statement:

```
there_is_something_worth_seeing
AND
I_have_enough_money
AND
NOT(I_am_busy)
```

I can work out whether it is true or false that I should go to the cinema. And since I can represent all of this electronically, using logic gates, I can build myself a natty little machine to take all the effort out of the decision. I can simply flick switches to denote the truth of each of the variables and then watch to see whether a lamp lights to tell me I should go.

Naturally, this would be a pretty stupid thing to want to do, and perhaps explains why nerds like me don't get out much. If that's as far

[1]If the oscillating electromagnetic field of a radio signal were first converted into a mechanical vibration, it would be possible in principle to build a complete radio receiver out of cogs and ratchets! Tuned amplification, rectification and demodulation are all feasible. The end product of a radio is already a mechanical vibration anyway – the movement of a loudspeaker cone.

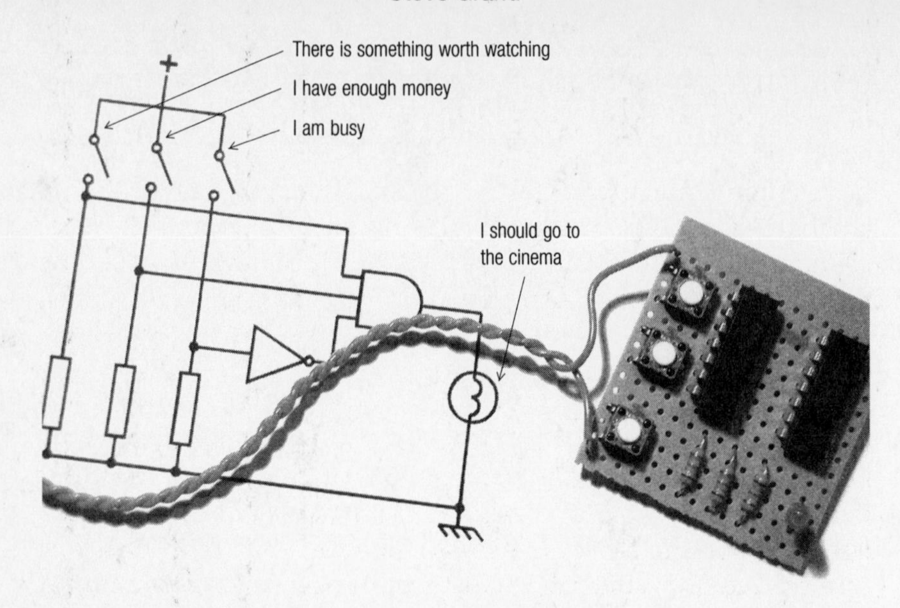

Figure 1 Deciding whether to go to the cinema, using logic circuits

as it went, the digital revolution might not have amounted to anything at all. But it turns out that by treating the ones and zeroes of a binary number as if they were truth-values, Boolean algebra can be used to perform binary arithmetic. It is an immutable logical fact that 1 + 1 = 10 in binary, and therefore this can be represented using electronic gates in essentially the same way that we can state the conditions under which I should go to see a movie. From such a tiny acorn the huge oak of the digital revolution has grown, although this is not the place to tell that story.

The only thing I will mention here is that digital logic comes in two flavours, just like clockwork with its linear cogs and non-linear ratchets. The simplest kinds of digital logic circuits are called combinational. The basic AND, OR and NOT gates are combinational, as are binary adders and other more complex structures. The characteristic of combinational logic is that the output is entirely determined by the current state of the input. Binary 1 + 1 always equals 10, no matter what the previous sum was. On the other hand *sequential* logic contains what amounts to a memory of the past. Now the output depends not only on the present state of the inputs but also on a memorised internal state, which itself is a product of previous inputs.

Perhaps the most obvious example of sequential logic is the binary counter that lies at the heart of a stopwatch. It remembers its previous output (a binary number) so that every time a new input pulse (a binary '1') arrives it can add the latter to the former to compute the next number in the sequence. Even though the input pulses are an absolutely identical series of 1s, the output is different every time, counting incrementally up from zero. So while simple combinational logic is a bit like a novel – no matter how many times you come back to it, you will always read the same story – sequential logic is the soap opera of the computational world: ever changing, often surprising, and sometimes totally baffling.

The reason I'm telling you all this is to introduce the question of how we might compute things using neurons. Neurons, just like cogs, transistors and logic gates, are basic computational elements. They don't do anything particularly startling on their own, but when you connect them together in the right ways you can make them do amazing things. After all, it was the arrangement of neurons in my head that was responsible for composing and producing this chapter.

But what do they actually *do*? Cogs compute simple ratios, transistors do something similar with electrical currents, and logic gates perform truth operations and binary arithmetic. So what do individual neurons do?

A neuron is just a cell, like any other cell in the body, and hence is an intricate machine made from millions of molecular moving parts. But unlike most cells it has a pronounced capacity to grow long thin tendrils called axons and dendrites (some axons are more than a metre long), and a highly developed ability to respond to electrical and chemical changes on its surface. None of this is unique to neurons; it's just that neurons are specialised for transmitting focused electrical signals over long distances. The detailed electrical, chemical and physical behaviour of a neuron is unbelievably complex, and contingent upon a huge variety of factors.

This is deeply unfair. Cogs might suffer from friction and inertia, and transistors aren't as conveniently linear as we would like, but it is pretty easy to abstract them into idealised forms that are convenient to think about. Digital logic circuits go to great lengths to make the messy realities of electronics irrelevant to their operation, and it is in the nature of digital systems that they behave (most of the time) precisely as their idealised abstractions would predict. But this is not so for neurons. They are far more complex and intricate than a cog. So which particular aspects of a neuron's behaviour are important for its role in computation, and which are merely a consequence of the fact that it is made from biological materials?

Scientists abhor messy complexity, and do their utmost to simplify and idealise things so that they may be understood. As you might expect, therefore, the connectionists very quickly latched on to an idealised description of the neuron, which they hoped might encapsulate all the important things while leaving out the messy irrelevancies. Most neural networks assume that a neuron is a device that carries out a sum-of-products calculation on its inputs. In other words, each neuron has a number of input wires called dendrites, each of which provides it with a numerical signal from another neuron or from the outside world. Between the output (axon) of one neuron and the input of the next is a junction called a synapse, which in real neurons is the site of a complex transformation from an electrical signal to a chemical one and back into an electrical one again. In the idealised neuron a synapse is simply assumed to multiply the input signal by a certain fractional amount called the synaptic weight. All these attenuated input signals are then added together in the cell body (the part between the dendrites and the axon) and then sent along a single axon, finally to arrive at another synapse. So the output of an idealised neuron is the sum of all its input signal strengths, after being multiplied by their respective synaptic weights. Often the output is slightly modified to make it harder for the neural signals to get awkwardly out of range, but essentially that's all there is to it.

People have tried to join these idealised neurons (or variations on the theme) together into networks in many different ways, with varying degrees of success. The most widely used method is so well known compared to the others, that when neural networks come up in conversation (as they do at all the best parties), this is what most people assume you mean by the term. In this arrangement, the idealised neurons are arrayed in three layers, with each neuron being connected to every one of the neurons in the layer beneath. Already we can see how this has nothing to do with the brain, because the number of connections in such a network increases exponentially as more neurons are added. Practical networks of this type often have no more than a few dozen neurons, and scaling this up to the billions of neurons in the human brain is clearly both a bit of a faint hope and biologically

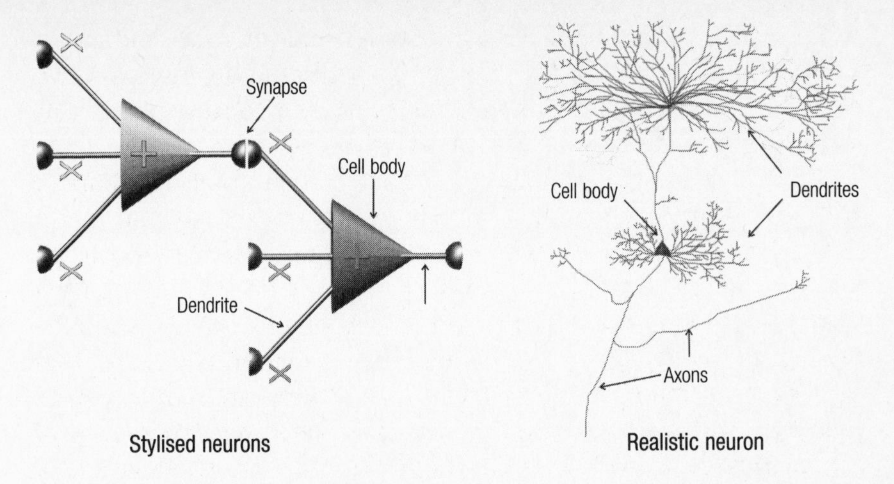

Figure 2 A pair of stylised 'connectionist' neurons, compared with a (simplified) real neuron from the cerebral cortex

completely unrealistic. Nevertheless, three-layer networks like this can be trained to recognise patterns in their input data. They do this by being exposed to a set of known input patterns, to determine what pattern of outputs their initial (usually random) synaptic weights produce. This output is then compared with the correct answer and the synaptic weights are adjusted so that next time the result is closer to the intended one. Gradually the network settles down to a fairly good approximation of the right answers for the known patterns, and then, with luck and a following wind, goes on to produce the right answers when presented with unfamiliar examples.

This is all very impressive I'm sure, but if this idealisation is enough to explain how real neurons compute then I'm a cartoon character. Yet variations on this theme have managed to serve connectionist researchers passably well for a long time, in a learning-to-jump-at-the-moon sort of a way. In fact you can do a surprising amount of stuff using only addition and multiplication, as any mathematician would be delighted to demonstrate. The particular operation of sum-of-products is actually so useful that a special class of microprocessor, known as a digital signal processor (see chapter 6), goes to the trouble of implementing the operation as a single, highly efficient instruction known as MAC (for Multiply and ACcumulate).

If you want to know more about how traditional neural networks work, then you'll have to read a specialised book on the topic. And good luck to you too, since the subject can get pretty rarefied and esoteric very quickly. But what does all this idealisation have to do with brains anyway? Well not much, if the truth were told. I'd go so far as to say that 'absolutely zilch' is only a minor exaggeration. Can I build Lucy using a three-layer, sum-of-products network? Heck no! There are so many things wrong with this kind of abstraction that I hardly know where to begin, but for the purposes of this section of the book I'd like to draw two parallels with digital logic.

The first of these I shall merely mention now and then come back to in the next chapter. The basic elements of logic circuits are AND, OR and NOT gates, but these gates are not indivisible atomic entities; they are themselves made from small circuits of transistors. A typical two-input AND gate is made from six transistors arranged in a specific way. Moreover, even the individual logic gates are pretty dull, and life gets much more interesting when they are collected together into a small number of even higher-level building block types. When we design logic circuitry we sometimes scatter the odd individual AND gate around but most of the time we use larger, composite structures as our base-level tools, such as flip-flops, counters and shift registers. Obviously brains are made from neurons, just as computers are made from transistors, but the designers of computers rarely think in terms of modules this small. All their theoretical and conceptual activity takes place at a higher level of description. Perhaps the designers of artificial brains need to think this way too. Perhaps the designer of our own brains, natural selection, actually did deal in units much larger than a single neuron. The standard approach to neural networks just takes it for granted that the basic computational element is the neuron, and in the vast majority of cases assumes that all neurons are alike. This is, I think, a serious mistake.

The second parallel with digital logic involves the distinction between combinational and sequential forms of logic. Remember that sequential logic blocks maintain internal states that act as memories of past experience, so that their behaviour depends not just on the present but also on their previous history. Sequential logic is *much* more interesting than combinational logic, if distressingly difficult to analyse. By contrast, the traditional style of neural network has no internal state, and always produces the same answer, given the same inputs.

'Hang on,' you might say, 'what about the way that synaptic weights are adjusted as the network learns? Isn't this an example of a stored internal state?' I'm glad you pointed that out (which of course means that I hoped you wouldn't notice, since now I have to explain myself). Changing the synaptic weights during learning does indeed involve feedback, and does mean that the behaviour of the network at any moment depends on its entire past history. But this happens on comparatively long timescales. It is more like an occasional redesign of the circuit. The kind of internal state I'm talking about has an influence on a moment-to-moment basis, and gives rise to complex *dynamics* in the network.

To produce such dynamic behaviour, some of the neurons need to be turned round the 'wrong' way, forming loops that feed some of the output back into the mix. Such a network is called *recurrent*, and recurrent networks have startlingly more interesting properties than simple feed-forward ones. There are a number of standard schemes for recurrent networks in existence, but they do tend to be quite restrained in their operation, because recurrency really gives the theoreticians nightmares.

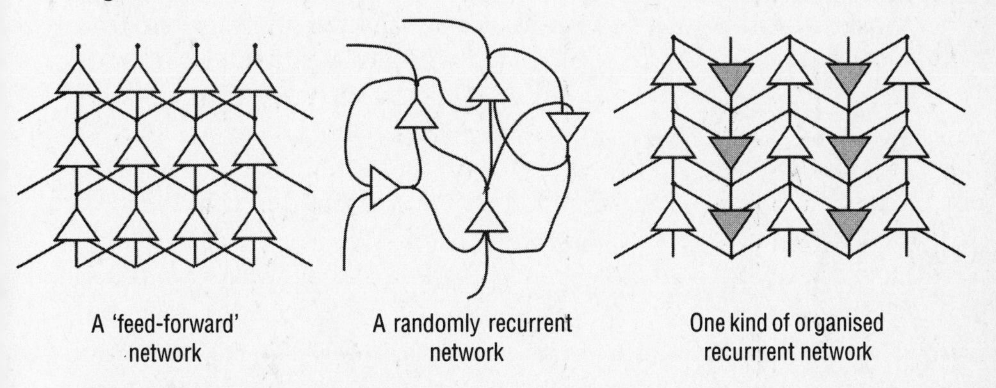

Figure 3 Some network topologies

In truth, the universe is stuffed full of highly dynamic systems involving feedback like this. Think of a simple conversation between two people, for example, or the stock market, or a thunderstorm. Because we have no effective mathematics for dealing with such monsters, more often than not we pretend that they don't exist. Complexity theory is a modern method for trying to tame the beast, but although this can tell us the *kind* of thing that any given network of interactions might get up to, it tells us nothing about what it will *actually* do, and precious little that will help us design a system to do what we want it to do. We know that the brain is an incredibly recurrent structure at almost every level, but we generally prefer to close our eyes to this fact and hope it will go away. We simply don't have the analytical tools.

But it can't be ignored. Sequential logic is not mildly different from combinational logic; it is radically different. Equally, feed-forward neural networks are not a convenient approximation to recurrent ones; they have almost nothing in common. What are we to do in the face of such turbulence? One possible answer, short of baling out of the plane in desperation, is to stop meddling with the controls and let an autopilot fly the damned thing. In other words, let something design the network that doesn't need to use analysis and theory to get what it wants. One such thing is evolution. Highly recurrent networks have unpredictable and bizarre properties, but evolution doesn't care one jot. If it works, evolution keeps it and improves on it; if it doesn't, the Grim Reaper tears up the blueprints. Evolution doesn't have to write papers on why something works; it simply has to know *if* it works.

Since a computer can represent virtual objects, such as neurons, and these objects can be plugged together into networks using nothing more than lists of numbers, it is quite easy to build simple virtual creatures (or even real robots) whose neural network 'brains' are initially connected up at random and then evolve into more meaningful structures through natural selection. A common post-graduate project is to develop a virtual creature whose motors and sensors are connected to a small network (maybe sixteen neurons) in an arrangement specified by a string of numbers that play the part of the creature's genes. Most randomly wired networks are unable to do anything useful at all, but if many such creatures are 'bred' at random and tested in a virtual arena for their ability to seek out food, avoid hitting the walls, or perhaps chase each other, then one or two may have a slight advantage over the others. Taking the genes for these mildly successful creatures' networks and making copies with slight modifications (mutations) creates a new generation of creatures, many of which will be as good at the task as their parents, while some may even be better. By repeated testing and breeding, allowing only the best from each generation to propagate their genes, the neural networks will gradually evolve the ability to tackle the task better and better. And nobody ever needs to know how it works.

Hey presto! It's like having a letter from your mum, excusing you from lessons. Suddenly you don't have to sit there in front of a blank sheet of paper, desperately trying to think up a recurrent neural network structure from scratch. Now you can get on and do real experiments. Let nature take the strain, and then try to learn something from what evolves (or if necessary find a good excuse for why nothing useful evolved at all). This kind of reasoning has led in recent years to a whole new industry within the AI and biological communities. You could almost hear the sigh of relief as all these scientists, so frustrated by the limitations of the high jump, suddenly realised that there was a new way forward. At last the moon would be within reach, because now they had discovered how to pole-vault!

You may think that this is a cynical remark, and it is, just a bit. In fact, for those people brought up in a world filled with neat linear mathematics and hopelessly ill-conceived attempts to replicate human-level intelligence, evolutionary neural network research has been a genuine revelation and a breath of fresh air. For example, it has demonstrated that many kinds of behaviour, which previously seemed to require significant intelligence, can actually be replicated using very tiny neural circuits, and sometimes no neurons at all. It has shown that intelligent behaviour does not have to arise from any central master controller, but can emerge out of much simpler interactions that are distributed around a creature's nervous system and body. It has also alerted people to the fact that intelligence is not an isolated thing but arises out of the tightly coupled interaction between a creature and its environment.

But it is important not to replace one dogma with another. Remember that these evolved neural networks contain only a handful of neurons, wired together in a haphazard, unstructured way. And the inspiration for these systems is not drawn from human beings or even from rats, but from invertebrates, particularly insects. Unfortunately, the lessons one draws from a cockroach do not necessarily apply to a human being.

CHAPTER FOUR

Cockroaches to cortex

Oh dear. Things aren't going terribly well, are they?

As we have seen, I can't directly program Lucy's computers to act intelligently, because minds are not like computer programs. In any case, behaviour that has been explicitly programmed in is more or less the exact opposite of intelligence – when people behave as if they are blindly following rules we don't regard them as intelligent but call them zombies. What I need to do instead is to emulate the brain directly, using a neural network, so that intelligence arises as an emergent phenomenon that develops all by itself, through learning, rather than being hard-wired by me. But the conventional approaches to neural networks, using idealised neurons arranged in simple, usually non-recurrent ways, are hopelessly inadequate and have almost nothing in common with how real brains work. Evolution, it is true, can build messy, recurrent networks that have slightly more biological plausibility, but unless I'm willing to wait a while for them to evolve – say five hundred million years or so – I would be restricted to no more than a handful of neurons. It's a good thing I knew about these problems in advance, or I might start to get depressed!

Of course, I've been telling you all this bad news because I am trying to lead up to something. By building Lucy I'm taking a specific approach to AI that makes sense to me, after many years of thinking about these issues, but it may not make sense to you unless I point out some of the problems with the more traditional or obvious ways of doing things. I don't want to bore you with too much background, and I know I promised you robots and other glamorous and exciting things, but the homework is nearly over. Before I heat up the soldering iron, I just want to tackle one more misconception that will help to explain exactly what I'm looking for.

Most academics of my acquaintance are encouraging, helpful and positive about what I am doing, even if they secretly think that I'm insane. After all, I'm only pursuing an innocent hobby in my own sweet way, at my own expense, and doing nobody any harm. But I do occasionally upset people nonetheless. Just recently I managed to trigger a torrent of personal abuse from a certain doctoral student, whom I shall call Mr X to preserve his anonymity. I suppose it was partly my own fault for giving a really shallow and lightweight seminar at his university, when I should have realised that what I'm doing upsets the ideological applecart in certain quarters and needed addressing more deeply.

The main drift of Mr X's criticism was that I shouldn't go around attempting to model something as complex as the mammalian brain when we have barely a clue about how the brains of very much simpler animals work. He reeled off examples of the impressively complex behaviour of tiny insects, and accused me of rashly trying to sidestep millions of years of evolution. Never mind the fact that insects are actually our distant cousins, not our genetic ancestors, and are therefore just as highly evolved as we are: on the surface Mr X does seem to have a point. How can anyone hope to understand a brain as complex as a human being's, when we don't even understand the three hundred or so neurons that control a microscopic roundworm, one of the simplest multi-cellular animals in existence?

First of all, let me make it clear that I'm not that stupid. At best I can hope to give Lucy a few tens of thousands of neurons, which is more than a million times smaller than a human brain. I don't really believe that she will ever sit up and demand to be taught calculus. She is just a test bed for ideas – a vehicle for thinking with, not a goal in herself – and if she even manages to figure out for herself how to reach out and touch something she has seen then I shall fall off my chair in astonishment. But I have never claimed to be in search of human-*level* intelligence; what I am trying to understand is a specific aspect of human-*style* intelligence (more accurately, mammal-style intelligence), and this is not the same thing at all.

For that matter, I think the claim that human brains are vastly more complex than roundworm brains is nowhere near as defensible as it sounds. Measured gram for gram I think it's actually possible that our brains are *simpler* than those of some invertebrates. Before you take affront at this insult to your humanity and throw this book across the room, let me try to explain.

In 1984 the neuroscientist and cybernetician Valentino Braitenberg published a small but highly influential book *Vehicles: experiments in synthetic psychology* (MIT Press), in which he described designs for some hypothetical artificial creatures that could show lifelike behaviour. For example, one of the simplest of these creatures can 'intelligently' navigate towards a light. It has two photocells connected to a pair of motors driving its wheels. The photocells are connected by wires (in a real animal these would be neurons), which cross over, so that the left light sensor controls the speed of the right-hand wheel and vice versa. If the source of light lies to the left of the creature, the left photocell will produce more current, which speeds up the right-hand motor and makes the creature turn left, until the light source is centralised. Braitenberg went on from this to outline an increasingly sophisticated (but by comparison with our intuitive expectations, incredibly simple) hierarchy of creatures, with the more advanced ones able to show such apparently complex states as 'fear' and 'love'.

Never mind the vexing question of whether these simple creatures would *really* exhibit love and fear, rather than just behave *as if* they were in love or fearful, Braitenberg did successfully demonstrate several very important things. First, that it is often easier to design a mechanism from scratch to do something, than it is to figure out how nature achieves the same effect – synthesis can be more powerful than analysis. Second, he showed that natural systems might not always be as complex as they seem. At a time when AI was in a state of denial about its inability to discover a single, over-arching 'algorithm of thought', Braitenberg's conclusions were a comforting solution to many. Perhaps no such thing was needed after all; perhaps all apparently intelligent action could be accounted for by simple, specialised structures.

For my part I strongly agree with his first point (which is why I'm building a robot, rather than dissecting a monkey), but I have some issues with the way people interpret his latter conclusion. When I say that human brains may be simpler in principle than some invertebrate brains, this is not because I think that love and hate are much less sophisticated than we thought, but because I believe that human brains contain an important simplifying factor that the simplest creatures do not. Braitenberg's observations do not predict or acknowledge this important fact and so should not be taken too far.

Braitenberg's logic seems to be supported by the rule of thumb known as Morgan's Canon (set out by the psychologist C. L. Morgan), which entreats people not to explain animal behaviour by recourse to 'higher psychical faculties' when a simpler, more straightforward explanation would account for the facts. In itself this sounds like good advice and seems very much like a law of parsimony along the lines of Occam's Razor: if we are faced with two possible explanations for something, the simpler of the two is to be preferred. Morgan's Canon perhaps lies at the heart of much of the behaviouristic zealotry that seeks to reduce all intelligence to specialised modules. Yet the Canon is not equivalent to Occam's Razor at all, and the reasoning is flawed. In any one instance, finding a simple, specialised explanation for a given behaviour is always more parsimonious than invoking some fancy mental apparatus. Thus far Occam and Morgan agree. But suppose we need to explain a million such behaviours. Morgan's Canon would now seem to imply that we should invoke a million simple explanations, one for each behaviour. But Occam's Razor directs us to find the simplest explanation overall, and one complex but general-purpose solution is likely to be far more parsimonious than the sum of a million special cases. Our brains seem to fit somewhere between these two extremes, because evolution has discovered some clever tricks that maximise computational power while minimising complexity.

Up to a point it is possible to build structures in the way that Braitenberg hypothesised in his book and numerous evolutionary neural network experiments have emulated since, but beyond a certain threshold of complexity the rules tend to change dramatically. There is an interesting parallel here with the history of electronics.

Have you ever looked inside an old valve (vacuum tube) radiogram or television set? If so, you would have been forgiven for thinking that the components had been set in place by throwing them into the air and soldering them where they fell, with the resistors and capacitors strung around from solder tag to solder tag like washing lines in a refugee camp. In a simple circuit like a radio set, or a very primitive creature, this is the logical thing to do. Each component, each neuron, has its own specific function, and there is very little repetition, so no organised structure is needed. I call such networks '*ad hoc*', since they seem to have been thrown together at the last moment. If you want to add new facilities to such a system, you can bolt on a few more components and maybe replace some of the existing ones, but in doing so the tangle gets more and more complicated.

Figure 4 Valve radiogram

If you keep adding new features, sooner or later you'll find that things get into a serious mess. Every time you add something new, you are likely to destabilise the rest of the mechanism, because everything's connected to everything else. Maybe you'll eventually decide to go back to square one and redesign the whole circuit from scratch, but this is something that evolution can't do, of course, since every generation must build upon a previously complete and viable design. Continuing to adapt and improve an existing *ad hoc* structure gets increasingly difficult as it becomes more complicated, and sooner or later you'll have to give up. Evolution will effectively give up too, as the rate at which it discovers improvements that don't damage the existing system falls steadily towards zero. But if you're a sensitive designer you'll probably soon make a discovery. You'll realise that a lot of the things you add to the system are actually quite similar to each other, and a single, more generalised module can be designed, which can be adapted to many tasks.

In electronics, the most obvious parallel was the development of the operational amplifier. Op-amps are generalised amplifiers, used as building blocks in a huge variety of circuits. In a radio set, for example, an oscillator circuit and a mixer circuit turn the high-frequency radio signal into a fixed lower frequency, this fixed intermediate frequency is efficiently amplified by a special amplifier, the audio component of the signal is separated out and amplified by a power amp and the volume of the signal is kept within bounds by an automatic gain control. Each of these stages used to be designed in its own unique way, but people gradually realised that they were all variations on the same theme; that a single, general-purpose building block could be designed, which would allow all of these circuits (and much more besides) to be built quickly and easily. In the days of discrete transistors, there wasn't much justification for doing this, but with the invention of the integrated circuit it became far more sensible to design one general-purpose building block and reconfigure it for each specific use, than to design each element individually, and so the ubiquitous op-amp was born.

In nature, the need to construct repeated instances of identical neural circuits arises whenever you have sensors or motor systems made from many parts operating in parallel. An insect's compound eye, for example, is made from many light-detecting units, called *ommatidia*. Each unit needs to perform the same neural processing on its input signal, and integrate this signal with those of its neighbours in a consistent way; hence each ommatidium's neural circuit is an instance of the same basic building block. In itself this is not quite analogous to the electronic op-amp, because although the same design of neural circuit is used many times over, it is being used in exactly the same way in each case. The thing about op-amps is that they embody something much more generalised, which can be put to many, quite different uses.

Now the question arises, did animals such as the insects evolve repeatable structures that could be applied to several superficially different but fundamentally similar tasks? Does the task of processing and integrating signals from a compound eye bear a sufficiently strong similarity to the task of processing and integrating smell signals from an antenna, for example? I don't know the answer to this in the case of insects, but it seems quite possible. If nature managed to discover a small neural circuit with the versatility of an op-amp, she would be sure to use it.

Not only would this modularisation simplify what would otherwise be a complete mess of *ad hoc* wiring, but crucially it would also allow one processing mechanism to evolve quickly and easily from another. The general principle of duplication followed by modification is one of the key ways in which evolution simplifies the problem of adapting to new circumstances. Many animals, particularly the arthropods (which includes the insects), have strongly segmented bodies. Each segment is derived from the same basic floor plan: a pair of legs, a pair of gills, a section of gut, a pair of neuron bundles and so on. But different segments become modified and specialised in different ways. It is easier to duplicate something that already works and modify it for a different task than it is to evolve a whole new mechanism from scratch. For example, insect wings evolved from either the pre-existing gill plates of a body segment or a part of the leg structure, and presumably the neural circuits that drive them are just slight modifications of the circuits that originally controlled leg movements.

If the local neural circuits in each segment are sufficiently flexible that they can easily be adapted for new uses, then perhaps the central ones in an insect's brain can too, especially since each of the pairs of ganglia (nerve bundles) that constitute its brain are probably derived from those of individual body segments that have become fused together to make up the other structures of the head. One pair of these ganglia processes signals from the eyes, another processes signals from the antennae, and a third combines the results from the first two. If they are all evolved from the same type of original structure, do they all work on the same basic principles? Are they merely stocks of neurons that end up being wired into radically different circuits in each case, or is there a central principle akin to that of the op-amp, which is flexible enough to prevent evolution from having to start from scratch every time?

As I say, I don't know the answer to this question, and to be honest I don't care all that much. Insects may be the most successful class of animal on the planet, but they are an example of the intelligence of *evolution*, and evolution is not what I'm studying. Individually, they aren't the most sparkling wits in existence, yet over quite short evolutionary time-spans insects have managed to invade almost every ecological niche. Taken as a whole class, insects can perform hard-wired and species-specific but still rather impressive feats, such as landing on a petal, navigating back to their burrow or giving comprehensive restaurant reviews via the medium of dance (in the case of honey-bees). It seems to me quite possible that they achieve such evolutionary flexibility by having access to a highly generalised, repetitive and easily modified building block structure in their nervous systems.

Whereas insects have responded to ecological challenge by rapidly evolving into new niches, mammals, and especially scavenging omnivores like us, have dealt with it very differently. When our environment changes (perhaps through a change in climate or the drying up of a food source) we don't sit around waiting for the survival of the fittest to solve the problem, we solve it ourselves, by thinking inventively and learning new facts and skills. If anything implies the existence of a generalised, reconfigurable building block structure, it is this ability to learn to do new things.

Let me return briefly to the history of electronics, since we now need a better analogy than the op-amp. In digital electronics, one of the most profound innovations of recent times has been the invention of the Programmable Logic Array. PLAs are very large arrays of extremely generalised logic blocks. The precise details vary, but in each type of PLA it is possible to configure individual logic blocks to perform any one of a wide variety of tasks that were previously implemented using specialised combinations of AND, OR and NOT gates. When you come across a large black slab of a chip inside a computer or other piece of domestic equipment, if it isn't a microprocessor the chances are it's a PLA. Designers use PLAs to produce very densely packed and complex logic circuits very quickly – the same type of chip might be used to control the input and output peripherals of a PC, the motors and solenoids of a video recorder or the keyboard of a mobile phone, simply by configuring it for that purpose. To configure a PLA for a specific task, the chip is 'programmed' by making or breaking links between and within the thousands of logic cells. The basic design of each block doesn't need to be changed, only the fine configuration details and the layout of the connections between blocks.

The basic element of a PLA is highly flexible and extremely easy to adapt for new tasks, unlike the fixed functions of AND, OR and NOT gates, and these elements are present in huge arrays. In practice, each element is too primitive to do very much by itself, and PLAs are used in place of discrete logic gates to produce relatively small numbers of complex circuits, rather than the very large numbers of moderately simple ones that would be needed to process the signals from an insect's eyes. But they make a good analogy. It isn't difficult to imagine that a similar fundamental processing element could be made from neurons, which, after minor reconfiguration by evolution, could process either the visual signals from an eye or the olfactory signals from an antenna. Perhaps it would even be possible for some of the fine details of this configuration to alter during an insect's lifetime, leading to a limited kind of learning.

In general, insects can learn by degree but not by kind: an individual insect can learn its way back to its own nest using landmarks that are unique to its particular circumstances, but it can't choose to do anything other than navigate with this circuitry, since most of the configuration details were set in place genetically, as a result of evolution. The main evolutionary advantage of a hypothetical PLA-like structure like this is that such gross changes can be made much more quickly and easily by evolution than would be the case if the basic element were as primitive as a single neuron. But suppose an even more powerful building block structure happened to evolve, which could rewire *itself* to perform a range of tasks. Instead of being wired up by genes, just suppose that this array of building blocks could become wired up *by the very information that is flowing into it*. This would be a miracle indeed.

Not only would such a magical machine be far less reliant on the hit-and-miss nature of evolution for discovering the right configurations to perform certain tasks, it would also be more robust and able to reconfigure itself to cope with changes in a creature's body (such as damage to a limb) or environment (such as a change in food availability). Not only that, but it would be capable of learning whole new tricks: not just minor adjustments to a hard-wired mechanism that evolved for a specific task, but the invention of completely new facilities – like the ability to play the piano, for instance.

You may think that all this is supposition, but it isn't. We already know that our brains contain many structures that have just such a generalised array-like arrangement. Some of these may be wired up by evolution (perhaps indirectly, see chapter 15), but some, most notably the cerebral cortex, which is responsible for our most sophisticated conscious and unconscious learned behaviour, are capable of wiring *themselves* up to perform a range of specialised tasks.

The cortex is the thin, pinkish-grey 'rind' that lies on the surface of our pair of cerebral hemispheres, the largest and most obvious part of our brain. Essentially, cortex is cortex is cortex: no matter where you look, the basic structure is the same – not merely a homogeneous mass of identical neurons, but a highly organised, repetitive array of neural circuits. Of course, it does vary in detail from place to place, since by the time we are born, or soon after, it has wired itself up to perform different tasks in different regions. None the less, we can tell that it is an array of fundamentally uniform and general-purpose units, because it is possible to cause a region of cortex to configure itself in a very different way (making what would normally be a visual region start to look like a piece of auditory cortex, for example) simply by surgically swapping over the bundle of nerve fibres that feed it input signals.

So, we seem to have at least three possible fundamental neural architectures in nature. The simplest *ad hoc* networks suffice for very primitive creatures, and in these the basic building block is the neuron, plain and simple. However, beyond a certain level of complexity, nature uses something roughly similar to the Programmable Logic Array (although not using digital logic, I hasten to add). In this case, the building blocks are more than individual neurons; they are repeated *circuits* of neurons, easily capable of being reconfigured by evolution to perform somewhat different tasks. The third level involves a similar kind of high-level building block, but with the amazing capacity to configure itself during a creature's own lifetime.

There is also, I think, a fourth level, present in at least humans and other primates, which builds upon the underlying cerebral architecture and adds considerably to its power. The neural building blocks that make up the cerebral cortex of all mammals wire themselves into 'maps', or regions devoted to the processing of separate tasks. The first cortical stage of vision, for example, occurs in a map called V1, at the back of the brain, where a few million building blocks (several thousand million neurons in total) receive the individual signals coming from the optic nerve. The results from this processing then go on to form the inputs to several other maps, which have become specialised for other aspects of vision. Similarly, when we choose to move our muscles, this is driven by a pattern of nerve activity in a different map, called primary motor cortex, or M1. And so on. This arrangement of specialised (but largely self-organised) maps carries out much of our learned but unconscious activity. Yet we can also make voluntary decisions; we can initiate, alter or suppress our behaviour *at will*. We can even perform mental tasks for which no specialised and rigidly structured circuitry is likely to exist. To do this, it seems to me we may be switching information around among existing maps, altering the flow and hence making use of their specialisations in new ways.

For example, if I ask you to imagine the letter 'p' and tell me if it is the same shape as the letter 'd', you are likely to work out the answer by rotating a mental image of the 'p' and comparing it to a memory of 'd'. You clearly can't have a special mechanism in your brain for answering this precise question, or even for answering questions of this type. It seems more likely that you are making deliberate use of cortical maps that have already developed the ability to rotate and compare images *for other reasons*, presumably connected with normal unconscious visual processing. The difference is that you are somehow usurping this normal function and feeding in new information (in this case a mental image of 'p') with a specific aim in mind. Such a fourth, volitional level of brain structure possibly controls the flow of signals around groups of pre-existing self-configured maps in a highly flexible and general-purpose way, rather like a computer, even. This is something we call 'thinking'.

What I'm using Lucy to try and uncover is the clever neural trick that makes level three possible (and might one day provide insights into level four). What is the building block structure of the cerebral cortex? How does it adjust itself in response to input signals in such a way that it automatically does something useful with them?

Adherents to the older, symbolic AI movement think they already know what the fundamental building blocks of the mind are: they see them as symbol manipulation operations, such as those performed by a digital computer. The connectionists, including the more modern *ad hoc*, Braitenberg-inspired, invertebrate-style neural network researchers, disagree. They believe that *they* are the ones who know what the fundamental building block is: it's the neuron. I think they're both wrong. I think we know where to find the fundamental building block but we don't yet know how it works. I think the holy grail of AI lies in understanding the highly generalised properties of the basic element of the cerebral cortex – a circuit made from a few thousand neurons of several different types, wired up in a consistent and characteristic way. Individually these building blocks do something pretty simple and far from obviously related to the outward behaviour of an animal, but taken *en masse* they generate intelligence. Understand the structure of these circuits, what they do to the signals that enter them (from all directions) and what rules control their self-organisation, and we will be well on our way towards understanding the mind.

The human brain is amazingly subtle and complex, but it is not necessarily compl*icated*. If all the parts were different then the overall machine would indeed be complicated – this is how the human brain appears when people assume it is made from specialised *ad hoc* neural modules. But that is because they are using the wrong paradigm. In reality, many of the parts are identical, or at least essentially similar, like Lego bricks. There is only one thing you need to know about Lego bricks, no matter how many varieties they come in: the little bumps on the top of one brick fix into the holes underneath another. If the brains of mammals use a similar trick (and it can hardly be denied that this is the case) then there may be a relatively small number of things we need to know in order to understand them too. In fact they may be easier to understand than the tangled, unstructured brains of roundworms. After all, which job would you prefer: dismantling a large Lego house or unravelling last year's Christmas tree lights?

It seems to me that most of AI is therefore looking in the wrong direction and trying to use the wrong construction set. Strangely, most neuroscientists who study the cerebral cortex already know this: they know that the cortex is a highly structured, array-like, self-configuring machine, and they know that understanding the cortical architecture is the key to understanding the mind. But AI seems to pay them little attention, not least because the neuroscientists don't know how the system works either. Untangling a single cortical circuit is an incredibly difficult task, because it is made from so many extremely tiny neurons, connected up in a very dense tangle (each neuron has, on average, ten thousand connections to other neurons). Worse still, the behaviour of an individual circuit only really makes any sense in the context of a whole map, or group of maps, so you can't deduce the behaviour of the system from looking solely at one of its parts. This is something that I shall come back to later, because it requires us to think about machines in wholly unfamiliar ways.

All is not lost, however. As Braitenberg pointed out, it's often easier to use *synthesis* – to understand something by trying to build it – than to understand it by taking it apart. Nowhere is this more true than in a system whose important properties are emergent. If brains embody general principles in their structure, rather than masses of *ad hoc*, specialised solutions, then it is probably very much easier to work out what those principles are by starting with a blank sheet of paper and trying to build a brain from the bottom up, than it is to try dismantling one when you have no idea what you are looking for.

Given that there is some truth in the things I've said above, what I'm trying to discover (heroically futile though it may be) is the single clever engineering trick that lies hidden in the structure of the cortical building block and which enables it to compute so many apparently diverse things – after all, smelling, seeing, moving and deciding what to have for lunch don't on the face of it seem to have very much in common, yet we know they are all handled by the cortex (in conjunction with its associated structures). I also want to know what rules are being applied that cause this structure to wire itself up in response to input signals, in a way that automatically manages to do something useful with them. For inspiration, I know a little about what the wiring diagram actually looks like (although there are big gaps in our knowledge) and I'm familiar with some of the strange and often counter-intuitive behaviour of large ensembles of these circuits, either from other people's direct studies of the brain or from watching my own behaviour and that of other animals.

One other thing I need in order to try and make sense of all this information is a powerful test bed: a robot creature that has a variety of biologically realistic senses and motor systems that will enable me to find out if my ideas actually work. I don't think intelligence can be understood by looking at abstract, toy problems – if the environment and the problem are simplified too much, then intelligence isn't needed and doesn't have anything to act upon. Moreover, you can't bluff the laws of physics and you can't get away with vague ambiguities when you're explaining something to a computer. There isn't a programming language in the world that supports the command 'SORT OF', as in the instruction 'when the user presses this button, sort of work out the answer'. Building a real robot is therefore a great way to find out if my theories genuinely make sense. It also focuses my mind wonderfully and stops me taking anything for granted. If it works, then I am right; if it doesn't, then either I am wrong or I have missed something out (or my computer is broken, which is sometimes the easier option to accept).

It's time to meet Lucy.

PART TWO
Body

I once defined biology as 'the study of things that go "squish" when you step on them'. Generally speaking, robotics has more to do with things that go 'scrunch', or sometimes, if you leave your soldering iron in contact with them for too long, 'fizz . . .'

This section of the book has lots of scrunchy bits in it and a few occasions when things went fizz, but mostly it is about how to make the scrunchy bits behave much more like the squishy ones: how to turn the neat, prim, precise world of technology into the messy, distorted yet vastly more powerful world of biology. Every android needs a body, but the kind of body they are given has a very strong impact on their potential for developing true intelligence.

CHAPTER FIVE

Meat

My body is my temple. Admittedly, it's a shoddy, makeshift sort of temple: the kind of temple that the Ancient Britons might have made do with while they were waiting for Stonehenge to be finished, and then kept pigs in, but it's all I've got and it works pretty well, considering it's been sitting slumped in front of a computer for nigh on twenty-five years. Lucy, on the other hand, looks like what you get when you take an old vacuum cleaner apart to mend it, then give up and heap all the bits in a corner of the garage, in the vague hope that a magical race of pixie shoemakers might happen by one night, looking for a career move.

There seems to be a bit of a discrepancy here. In the comic books of my childhood, robots' and androids' bodies were lithely built from shiny titanium, housing lots of technical-looking gizmos that, in some unspecified way, gave them superhuman strength and agility. I don't seem to remember any thick cables dangling out of their backs, leading to huge stacks of batteries or bulky compressors. They didn't get bits of fluff jammed in their gear teeth, and they didn't have to plug themselves into the mains for a fourteen-hour recharge after a mere ten minutes of fighting with Dan Dare. I think perhaps the sci-fi writers were glossing over a few of the snags. In fact there seems to have been quite a conspiracy of silence about how one is actually supposed to build machines that can move with the grace, strength and precision of an animal.

Perhaps one of the closest things we presently have to a machine that combines human (in fact superhuman) strength with a passable level of agility is the mechanical digger. If you aren't too bothered by size and weight constraints, then I suppose you could use a diesel engine to run a compressor and drive hydraulic rams to control the moving parts of a robot, but I suspect there may be a little consumer resistance to the idea of robotic housemaids that weigh four tons and blunder around the house like a demented mammoth, spewing out noxious fumes and enough decibels to take out someone's eardrums. Unfortunately, when you try to scale things down to reasonable sizes, the situation seems to get worse, rather than better. If you've ever heard a model aircraft engine you'll know that a 5 cc petrol motor doesn't make any less noise than a five litre car engine; it just makes it at an excruciatingly higher pitch. Meanwhile, the relatively graceful dance of a digger's hydraulic arm becomes twitchy and uncontrollable in the absence of a large pendulous mass to damp down the motion, so hydraulics are not ideal for moving small mechanisms. And then there's the cost, of course. The huge (and lethal, on account of their utter lack of intelligence) robot arms that build motor cars are very precise and very powerful, but they aren't cheap by any stretch of the imagination.

Yet look at me. I'm no athlete, but I handle pretty well by comparison. I move smoothly and silently (if you ignore the occasional grunt), and I have enough raw power to lift several times my own weight. I can move rapidly enough to throw a stone, and precisely enough to thread a needle. And I can do all of this for hours at a time on the strength of a doughnut and a mug of coffee, burning energy at about the same rate as a light bulb.

The merest glance at the internal machinery that makes this possible will tell you that human technology still has an awful lot of catching up to do, while a slightly closer look will confirm that the way biological machines work is radically different to the way man-made machines do things. Our machines tend to be made from small numbers of large parts, for example, while biological machines are made from large numbers of small parts. A Boeing 747 is one of the most complex artificial machines in existence, and is built from something like a million separate items, which sounds pretty impressive when you realise that someone had to design them and someone else had to fit them together. But one small mouthful of pork chop can outdo that by a factor of a thousand. Most of the contents of that mouthful are the sorry remains of tiny muscle fibres, each one of which could, before it was cooked, use its molecular motors to enable it contract a little bit, exerting a minuscule force on the tissues around it. Yet multiply that tiny force by a billion and it becomes huge. Not only that, but there is a vast network of tiny blood vessels and nerve endings, bringing energy and control signals into intimate contact with the muscle fibres to give them a very rapid and finely honed response.

The materials technology available to biology is awesome. The contracting muscle fibres pull on tendons that have the tensile strength of steel, while these in turn pull on the incredibly lightweight, supple and sturdy structure of the bones. The whole system is sheathed in gossamer-thin membranes and lubricated by fluids and bearing surfaces that would put any grease or motor oil to shame. And there is one further, rather counter-intuitive difference. When we want to make machines that produce precise and fine movements, we build them from parts that are themselves precisely machined to very tight tolerances. We try everything we can to remove sloppiness from the joints and friction from the surfaces and we use rigid, hard materials that don't wear or deform. Nature, on the other hand, produces something squelchy and pliable, with no straight lines, extreme variability and a geometry that would give any control engineer who tried to predict its behaviour a nervous breakdown. While human-designed machines have to be precise so that there's a fair chance they'll move in the way we expect them to, nature relies instead on sensory feedback and an amazingly subtle control system to extract dynamic precision out of something that's fundamentally rather sloppy.

One day we will learn some of these tricks, I'm sure. The development of new materials and processes inspired by nature is called *biomimetics*. Velcro[1] is the best-known example of a biomimetic material, inspired by the structure of teasels and burrs. Researchers in this field are already trying to produce structures that are, at least superficially, similar to muscle tissue, for example by taking jelly-like substances that expand in the presence of a certain chemical, and constraining them in bundles of narrow, flexible tubes. As the jelly expands, it widens the tube and hence also shortens it, in the reverse fashion to the 'Chinese Handcuffs' trick, where a woven tube grips itself tighter around your two forefingers whenever you try to pull them out. If the chemical that causes the expansion can be supplied and removed rapidly enough, perhaps by converting it from one state to another using an electric current, then there is some hope that these structures will substitute for muscle in robotic applications of the future.

Nanotechnology is another great white hope for producing artificial

[1]Velcro® is a registered trademark of Velcro Industries BV

motive devices. It is possible to imagine ratchet-like structures made from millions of microscopic silicon 'prongs' that can contract and expand to order, dragging one surface along another in a manner reminiscent of a walking caterpillar, a method that is very closely allied to the way real muscles work. At an even finer level, molecular engineering can hope one day to replicate the very protein structures that drive real muscle contractions. However, to produce such devices in large numbers at low cost, we can hardly expect to build them a molecule at a time. Instead we will need to grow them: to use molecular tools to assemble protein structures in much the same way that DNA, RNA and enzymes assemble real muscle. If we get to that stage, it will be hard to tell whether we're talking about engineering or tissue culture. The distinctions between nature and artefact will have become so blurred as to be meaningless.

In the mean time, the existing methods of producing linear motion all have severe disadvantages, making them less than ideal for building robots. Not one of them is simultaneously smooth, controllable, powerful, energy-efficient and rapid. Wire made from an alloy called nitinol contracts when you put a current through it, but only slowly and weakly. Hydraulics and pneumatics are powerful and rapid, but not energy-efficient or easy to control. In a moment of desperation I even considered using steam power for Lucy, because boiling water is easy and cheap to make, the pressures are moderate and the equipment small and light. But in the end I decided that exploding boilers and scalding pipes might be a bit of a liability, and anyway I didn't have access to a machine shop for building the valves and cylinders.

Which brings me finally to electric motors. I hate electric motors with a passion, but they're all I have available. Several times in my life I've begun to build a robot in order to test some idea or another, but it was always the electric motors that defeated me in the end. In an ideal world, Lucy's limbs would be controlled by artificial muscles, but for now I'm stuck with using electric motors. Loathsome things! They are too fast, too weak, too hard to attach things to and too expensive. They go round and round instead of backwards and forwards and, worst of all, they swallow energy faster than a black hole.

The motors I eventually settled on are the sort used in model aircraft to move the control surfaces. They are relatively cheap, they come with a built-in gearbox, they don't draw too much current and they are just about strong enough to lift the weight of Lucy's arms and head. Sadly, Lucy has no legs at the moment. Lifting her entire body weight is far beyond the capacity of little motors like these, and larger motors draw so much current that her batteries would be flattened in seconds. Unless, of course, I gave her bigger batteries, which would mean more weight to carry, which means bigger motors ... You see why I hate electric motors?

It's a real shame, however, that Lucy (in her first incarnation), isn't able to move around, since mobility presents so many challenges and is a good test for my developing theories. Then why don't I just give her wheels, you might ask? Who said that robots have to have animal-like arms and legs?

You mean like the Daleks on the BBC's 'Dr Who' television series, perhaps? That race of ruthless cyborg megalomaniacs, who were hell-bent on taking control of every single planet in the galaxy, just as long as it didn't have any awkward steps to climb or rough ground to catch in their wheels? To aim for galactic domination when you could be foiled by a simple doorstep demonstrates a level of reckless ambition only surpassed by Eddie 'the Eagle' Edwards, the short-sighted, self-taught British ski jumper, from a land of little snow and absolutely no ski jumps, who valiantly and almost suicidally launched himself into last place in the 1988 Winter Olympics. Wheels (and skis) have serious disadvantages as practical means of locomotion and anyway, one of the biggest hurdles AI has to overcome is making robots that can live in the real world, so resorting to wheels is tantamount to cheating. Yes, I could have put Lucy in a wheelchair and let her come to terms with the challenges of being disabled, rather than leaving her completely stranded, but there's more to it than that.

Most research robots are not a bit android-like, much to the disappointment of the general public. Usually they are small, wheeled vehicles with an array of relatively simple sensors on them. They tend to run around in perfectly flat and unobstructed arenas, with high walls around the edge to prevent their sensors from seeing anything that they shouldn't. There are two main reasons why I'm not building Lucy this way. First, intelligence requires complexity. Imagine that someone has removed your brain from your body and placed it into a robot armed with nothing more than a couple of bump sensors and maybe a rudimentary form of sonar. You have no other sense; not touch, sight, smell, proprioception (the sense of joint and muscle position) or hearing, and the only actions available to you are to go left or right, forwards or backwards. How smart would you be after living this way for a while? How smart could you have become in the first place if you had been born like it? Life would be simultaneously too easy and too hard: too easy because switching a couple of motors on and off requires almost no intelligence or learning ability whatsoever, and too hard because the bottleneck now lies in your ability to interpret your environment using woefully inadequate sensors. You'd be lucky to end up with an intellectual capacity greater than the average thermostat. It seems to me that we should be deliberately making life difficult for ourselves and for our robots, rather than easier. When the problems are too easy or simplistic, intelligence is unlikely to be attainable, appropriate or even recognisable.

The other reason is subtler but I think it could be crucially important. The human eye is not remotely like a video camera, as I'll explain in more detail soon. Equally, animal muscles are not remotely like electric motors. Is this just an accident of nature, or is it important? In many respects, our eyes seem deeply inferior to cameras. If you could sample the signals leaving your own retina then you would hardly believe it was possible to see at all, but nevertheless, we do see very well indeed – better than any piece of technology yet created. Somehow our brains are at least able to make the best of things. But I think the issue runs deeper than this. Our brains and our sensory and motor systems have *co-evolved*. Brains have changed their structure over evolutionary time and the way they work will, we can presume, by now be beautifully matched to the way our sensors work. Meanwhile, our sensors will have evolved to capitalise on the way the brain processes information. The often bizarre kinds of signals entering and leaving our brains are therefore likely to be highly relevant to any understanding of how the brain works, so if I am to gain inspiration from natural brains it is vital that I emulate natural sensory and motor systems as much as possible. And of course, if I want to replicate the flexible and sophisticated behaviour of reptiles, birds and mammals, then there is little point giving my robot compound eyes and simple sensors like an ant. This doesn't seem to be widely appreciated in AI, and in many cases people's attempts to simplify the problem they are studying, by choosing sensors that seem to them more straightforward, structured and manageable, actually make things worse, since what seems simple to our minds is not necessarily simple to our brains.

This argument applies to motor systems just as much as it does to sensors. For all I know, the way the brain handles its motor output (and aside from controlling hormones and the immune system, flexing muscles is entirely what a brain is for), may rely very deeply on the way an animal's musculature works. Muscles are not things that you can tell to move to a specific position, like a motor; they are things that you tell to flex with a certain force. And they come in pairs or groups, with one muscle pulling against another. Perhaps if you were to connect raw electric motors directly to our central nervous system our brains would be able to learn to control them, but we can't rely on this, and all the neuroscientific evidence we have for how our existing motor systems work only applies to muscles.

It seems to me that Lucy needs reasonably realistic arm, head and eye movements (if sadly no legs), and they need to be controlled by reasonably realistic muscles. But all I actually had available was a selection of crummy electric motors, so one of my first tasks was to try and make these motors look and behave more like muscles, both in terms of their physical movement and in regard to how they appear to Lucy's nervous system.

The little model aircraft motors that I chose are of a kind known as servomotors. This may in fact be a misnomer, since a servomotor is strictly a small motor that controls the motion of a larger one. It could be that the name got applied to radio-controlled model servos because they were originally used to move the arm of a rheostat, which in turn controlled the power to a larger motor that drove the wheels of a toy car. Whatever its origin, the term is nowadays applied to any motor whose output is controlled using feedback from a built-in sensor, so that the shaft of the motor can be rotated automatically to a specified position. In the case of this particular variety, you supply the servo with a stream of digital pulses, whose durations are varied in proportion to where you want the servo's output shaft (which carries a rotating arm) to point. The circuitry inside the servo knows the present position of the arm because this is detectable via a small sensor mounted below the shaft. It compares this position with the desired position specified in terms of pulse width, and turns the motor until the actual position matches the desired one. All you have to do is tell the servo where you want its shaft to point by sending it some pulses, and it will put it there. If there is an external force resisting its motion, or something is trying to move it away from its target, the servo will respond automatically with an appropriate force to correct the error.

Figure 5 A radio-control servomotor

This is a quite neat mechanism with strong biological parallels, and the obvious thing for me to do would have been to attach a servo directly to each arm and head joint (Lucy needs at least a dozen joints, including moving her head, eyes and jaws). This way her limbs would arrange themselves in the desired positions just by sending an appropriate series of pulses to the servos. Unfortunately, there are several things wrong with this approach. Nothing is ever easy!

For one thing, suppose Lucy chooses to move her arm to a particular position but it hits an obstruction *en route*. I don't have a way of coating her body with hundreds of pressure sensors, so how would she know? Left to its own devices the servo will just crank up the power as much as it can, until the obstruction moves, the limb snaps or the motor grinds to a halt without reaching its goal. In our case we can feel whether our limbs are actually doing what we think they are doing by monitoring the stretch receptors in our muscles and joints (imagine weighing up a stone in your hand to see how heavy it is) and we can choose whether and how much to fight any resistance we encounter.

Tapping into the sensors inside the servos, so that Lucy's brain can detect the arm's actual position, seems like a plausible solution, but it doesn't help with another problem, which is that our limbs aren't always under tension and we can willingly allow them to relax and be moved by external forces. This could be important for Lucy because babies learn a lot about their world through having their limbs guided by a parent. If Lucy's servos resist all movement then I shall never be able to play pat-a-cake with her, and one of my big parental ambitions will have been foiled.

Coupled with this is the question of how the brain represents and deals with movement. Real limbs are controlled by pairs or groups of muscles pulling against each other, and somehow the nervous system is capable of juggling quite complex ballistic and controlled muscle contractions to achieve its ends, including letting all the muscles go completely limp when the brain simply doesn't care what position a limb is in (or when having free-swinging limbs provides some other advantage, as in walking). None of this bears much resemblance to simply specifying a servo position, and to avoid addressing these questions might cause me to miss something important.

Never mind; there is a solution. Instead of biologically accurate physical muscles, I've given Lucy *virtual* muscles, which, as far as her brain is concerned, behave in a fairly realistic way. Lucy's brain deals in terms of pairs of muscle tensions, but then a small computer converts this information into signals that drive a quite different physical arrangement. The way it works is like this: instead of the servos driving the joints directly, I made them pull on pairs of springs, and these springs then pull on the joint. Inside each joint is a separate sensor, which knows where the joint is actually pointing. If the limb has no resistance on it (including its own weight) then it will obviously always point in the same direction as the servo, but if there is a difference between these values I know that there must be an unequal tension on the two springs and hence an external force on the limb.

If Lucy chooses to relax both of the virtual muscles controlling any given joint, meaning that the limb should hang loosely, then what the physical system must do is keep moving the servo in such a way as to cancel out any differential tension on the springs. So, if I deliberately push on Lucy's arm the computer will detect the fact that the joint and the servo no longer face in the same direction, and command the servo to follow the motion. Lucy's brain will 'feel' her arm being moved (because the joint angle is changing in a way that

Figure 6 Lucy's arm 'muscles'

she didn't specify herself) and the arm itself will appear to move relatively freely because the servo automatically moves to prevent either of the springs tightening. But if Lucy chooses to resist my pressure on her arm, she can apply tension to one or both of her virtual muscles, which the computer interprets by creating an offset between the angle of the servo and the angle of the limb, causing a differential tension on the pair of springs.

This is not ideal, but it will have to do for now. One irritation is that, since the sensors in her joints aren't very stable under changing temperatures, and because my engineering is (to say the least) a bit sloppy, Lucy needs to recalibrate her limbs every time I switch her on. She does this by moving all her limbs to each end-stop in turn, and using the readings from the sensors to define the scale and offset values for future calculations. When she does this, she looks exactly like she's waking from sleep with a stretch and a yawn.

The only serious hitch I've found so far is that ordinary springs aren't really suitable. They lose their springiness too easily and so get longer and looser over time. They also spring back too fluidly and could do with being damped down to stop them overshooting their target and causing Lucy to flail around as if being attacked by bees. Instead of springs I've tried using bits of the rubber O-rings that you use to seal drainpipes. These have exactly the right kind of chewing-gum-like stretchiness and they don't overstretch. But piercing them with brass wire to fix them to the servo arms is rather hit-and-miss, and I don't think they'll last very long. It's all a bit weak and fragile, but so far I don't have any better ideas.

One day, I hope to be able to give Lucy a proper body, with complex joints and smooth, powerful muscles. In fact I plan to start work on a new body soon after I finish writing this book, using a compromise method that still involves electric motors but uses them to produce true linear motion with much less noise. For one of the things I didn't bank on was the dreadful noise that all these small plastic gearboxes make. When Lucy hears a sound, one of the things she needs to be able to do is turn her head to look at the source of it. Sadly, in practice, the noise her neck motors make is enough to stop her from hearing anything but her own creaking joints until she stops moving.

Such are the trials and tribulations of being an amateur roboticist. You start out with dreams of gleaming chrome, and you end up with a scruffy aluminium scarecrow. But even the best, most expensive technology available today is nowhere near good enough to produce the self-contained, agile, high-endurance bodies that my childhood comics promised. On the other hand, at least it makes me feel less inadequate about my own much-abused carcass.

CHAPTER SIX

Take your PIC

Don't you just adore the delightful creatures? My first brief and rather chaste encounter with them happened (I am now willing to relate) in 1970, at the innocent age of twelve. Needless to say, at that age you don't really understand what is going on. A mere *frisson,* perhaps: an amorphous and slightly illicit thrill somewhere around the loins. Like so many adolescents, I gained much of my early knowledge through old wives' tales, passed to me in a confidential whisper by a friend. Apparently, or so he told me, a sure-fire way to make a computer explode is to ask it the question 'If I told you that everything I say is a lie, would I be telling the truth?'

This information was pretty oblique, but nevertheless charged with useful facts for someone with an enquiring mind who was willing to read between the lines. For one thing, I deduced that the purpose of computers is to answer questions, presented to them in perfectly ordinary English. For another, it was clear that recursion – a snake that eats its own conceptual tail – was deeply upsetting to them, and was liable to make them destroy themselves in a fit of frustration. How glad I was to be able to benefit from my friend's greater experience of these things and save myself from embarrassing future *faux pas*.

Thereafter followed a period of healthy and largely contented celibacy. When I eventually had another tantalising and slightly more direct brush with a member of the mechanical sex in 1974, I didn't even recognise it for what it was – it bore so little resemblance to what I'd been led to expect. They say you always remember your first time, but for many of us I expect the details are a little hazy, if not pathologically suppressed. In my case I have only the faintest recollection of events. It started with a diagram in a magazine (as it so often does), and I remember that her name was Scamp. I was a little shocked – I had never seen anything quite like this before and it left me rather disoriented. Nevertheless, in these more enlightened times I see no harm in reproducing, for your edification, what I saw in the pages of *Practical Electronics*, some time in the summer of '74. It looked something like this:

```
005  LDA #017
007  STA !110
00A  BRA !042
```

Pardon? Any possible connection between this gobbledygook and an English-speaking, oracular, self-destructing computer simply passed me by at this point, and I didn't understand a word of what I saw. Not a single word. I was very unnerved, and I quickly came to the opinion that computing was not for me.

But in the end it all came fairly naturally, as it does to all of us, although I have to admit there was some fumbling involved. It happened one day when my girlfriend and I were alone, quietly punching some results from a physics experiment into our college's PDP8 minicomputer. One of us – I expect it was me – mistyped a number, and we discovered that pressing the Backspace key had no discernable effect. Up until this point our only keyboard experience was with a typewriter and somehow correction fluid didn't seem appropriate for a computer Teletype, but after an exhaustive search of the keyboard we came across a key marked ESCAPE. This seemed to sum up the way we were feeling about the problem so we pressed it, and this time something happened. The Teletype chirped into action and printed the words 'BREAK: 1100 >'. All now seemed to be well, so we kept on entering numbers and hitting the Enter key until we reached the end of the list, whereupon, instead of printing out the anticipated graph, the Teletype simply sat there, purring in anticipation of our next move. It had nothing more to say, and nor did we.

So we sought counselling, in the form of our physics lecturer, who took one look at the jumbled printout and sat down rather heavily, exhibiting a slight rolling of the eyeballs. It turned out that pressing Escape had stopped the program from running, and returned control to the command interpreter. From this point on, the computer interpreted every number we typed to be the line number of a program command, and a number followed by Enter meant that any existing command with that number should be deleted. How were we supposed to know that? Given that the computer had no disc drive, the lecturer had no option but to type in the entire program again from scratch. I was impressed. Clearly it was even easier to make computers explode than I'd been told!

This is a roundabout way of getting to the subject of microcontrollers. I don't suppose many people these days think that computers are English-speaking oracles of wisdom, but it may be the case that you haven't really thought much about the tiny computer chips called microcontrollers (or MCUs to their friends) and how they may be used as fundamental electronic building blocks. The first microprocessors (not quite the same thing as microcontrollers), such as the one I completely failed to understand in the pages of *Practical Electronics* in the mid-seventies, were never actually intended for use in computers at all, but were seen as components of a generalised control module for electronic circuits. The direct and honourable descendant of these early microprocessors is therefore the modern microcontroller, while the personal computer is actually something of a bastard child. Lucy relies on quite a number of these glorious little devices, so for those of you who are not familiar with them, allow me to introduce you now.

A normal computer system consists of a number of elements: a processor to do the work, some RAM memory in which to store the data and short-term program instructions, some non-volatile memory to hold more permanent software, such as an operating system and an assortment of peripheral devices and ports – gateways through which to interact with keyboards, displays, disc drives and other paraphernalia. The wonderful thing about a microcontroller is that it contains all of this stuff in a single chip – often a very small and cheap one. My first ever computer contained about a hundred chips and cost me (alright, it cost my girlfriend) £200. Today I can buy a computer that's more powerful than this and exists on a single chip the size of a postage stamp, for less than a fiver. Such microcontrollers are seriously cut-down and simplified computers, much less powerful than a modern PC, but they are very small, very cheap, and can run for hours on a torch battery.

This reduction in size and cost is not just a convenience, either. Sometimes quantitative changes can also lead to qualitative ones. After all, the only significant difference between a slide show and a movie is the speed at which the pictures are displayed, and yet movies have qualities that no slide show could ever approach, because they pass a crucial speed threshold, beyond which our brains interpret sequences of pictures as continuous motion. In the case of the microcontroller, the qualitative change from expensive computer to cheap MCU is rather more subtle but still important.

I will return to this in a moment, but first let us examine a typical and tremendously useful example of a microcontroller: the PIC16F876 produced by Microchip Inc. This example of the PIC range is easily my favourite MCU, because it is cheap enough that I can accidentally blow a few up without worrying unduly and it contains all the peripherals that any hot-blooded chap could desire. The processor at its core is a rather puny eight-bit device. In other words, it can count up to 255 with ease, but anything involving larger numbers requires significant extra effort and can quickly use up the available computer power. Simply multiplying two numbers together may involve dozens of instructions and a considerable amount of computer time. The way the processor is designed allows it to execute one instruction every four ticks of the in-built clock that synchronises the internal activities of any computer. Since the clock can run at speeds of up to twenty megahertz, this means the PIC can execute five million 8-bit operations per second. If you find this impressive (and it impresses the hell out of me, because I can't do arithmetic anything like that quickly), just bear in mind that a modern desktop PC has a clock that runs at up to two *billion* ticks per second, with a processor that can usually execute several 32-bit or 64-bit operations per tick.

The PIC's processor core may not be up to much, and compared to the hundred or more megabytes of memory in a PC, its complement of 368 bytes (yes, bytes) of volatile RAM and 8K of non-volatile program memory seems a little constricted, to say the least. In compensation, the 16F876 has lots of other goodies to play with, some of which will turn up in the following chapters. For example, it has an on-chip analogue-to-digital converter, with five input channels and the capability to sample input voltages several thousand times a second. Analogue-to-digital converters (ADCs) are immensely useful things that can convert a voltage into a binary number, and therefore allow a digital computer to interact with an essentially analogue world. An ADC can be used to monitor the position of a joystick, the temperature of a room or the rapid changes in signal strength produced by a sound, for example.

The converse of an ADC is a DAC – a digital-to-analogue converter, which, as its name suggests, allows the computer to convert a series of binary numbers into a more-or-less continuously variable voltage, with which to control external equipment such as an electric motor, or perhaps to generate a sound. Sadly, the PIC16F876 doesn't have a DAC, but there are ways you can get around this as long as speed is not a big issue. Notice that a microcontroller with a fast ADC and a DAC can turn continuously varying signals into numbers, perform calculations on those numbers and then convert the result back into an analogue signal again. This is the basis of a very powerful set of techniques called digital signal processing, which we will meet again later, when I describe Lucy's hearing and voice.

DACs and ADCs send and receive analogue signals (those which can vary continuously over a range). The equivalent for binary digital signals (things which are either on or off) is called an input/output port. The PIC has several I/O ports, each consisting of up to eight signal lines that can either sense the level of an input signal (near 0 volts is considered 'off' and near 5 volts is 'on'), or produce a signal that can be switched between 0 V and 5 V at will by a program instruction. Such ports can be used to detect whether one or more switches are open or closed, can switch one or more devices on or off, can exchange whole binary numbers with another device or be used in a variety of other ingenious ways (such as sending servo pulses to Lucy's muscles).

Sometimes computers exchange binary numbers with other equipment one bit at a time, instead of sending all eight bits in a byte at the same time. This is slower but it greatly reduces the number of wires needed to connect devices together, and is often more robust to electrical interference and cable length. The PIC has several ways to do this, using a number of standard protocols for defining the number format and control signals. One such standard is RS-232, which will be familiar to anyone who uses a modem or other serial device connected to a PC. As well as RS-232, which can be used to connect circuits containing PICs directly to a PC (via a converter chip that generates the necessary voltages), the 16F876 also speaks another common serial language, I^2C. This was designed to allow microcontrollers to talk to devices such as LCD displays or external ADCs without needing lots of signal wires, but interestingly it can also be used to allow MCUs to talk to each other, opening up possibilities for developing multiprocessor systems.

One particularly nice feature of this range of PICs is that they are in-circuit programmable, which means that you can program and reprogram them while they are soldered into the application circuit. This makes program development and debugging much easier than having to remove the chip and plug it into a separate programmer, or even throw it away whenever you make a mistake and program a fresh one, as you must with some types. PICs can be programmed using the C programming language, or in their native (and quite arcane) assembly language. Assembly language programming is a must if you need the most efficient code, as you almost always do.

So, a PIC is a complete computer on a chip. However this doesn't mean we need necessarily think of it as a computer. Remember what I said in chapter 2 about computers being places to build things, as well as devices for performing processes? Inside the cyberspace of a single PC you can build yourself a complete construction kit of tiny virtual machines, with any properties you like, and then use these to build something larger, whose behaviour emerges from the interactions between these virtual machines. A microcontroller can't easily implement large populations of virtual machines, but in itself it does conform closely to Alan Turing's concept of a universal machine – you can program it to emulate almost any single machine you like. This has quite a different feel about it to using an MCU as a sort of cut-down computer, controlling operations in a centralised way. Microcontrollers are so cheap and so flexible that you can often treat them as fundamental electronic building blocks – small objects whose internal programs are used to give them interesting external properties, rather than using them as a centralised 'brain' to control the entire system. This has a nice biological ring to it, and mirrors the populations of cells in the body, or neural circuits in the brain.

In principle, I could even make Lucy's brain from a large array of microcontrollers, rather than a single powerful computer. Each MCU could emulate the behaviour of one of the tiny neural circuits I talked about in chapter 4, and send signals to many other such circuits using, say, one of its serial ports. In practice this is impossible, since I'd need tens of thousands of microcontroller chips, which is more than my bank balance (and Lucy's batteries) could stand. But it illustrates how microcontrollers are qualitatively different from larger computers: instead of thinking of them as centralised master controllers, they can be used as highly flexible configurable building blocks. This is a bottom-up viewpoint instead of the more traditional top-down one.

More to the point, I can use microcontrollers as a means of turning hard-edged electronic sensors and motors into devices with a much more biological character. Lucy may have a television camera for an eye, but by attaching a suitably programmed MCU chip to it, I can make it produce signals that are much more like those from the retina. Similarly, a microphone can be turned into something more like the cochlea of the ear, and a loudspeaker can be made to look like a complete vocal tract, with lungs, vocal cords and the resonant cavities of the mouth and throat. And, of course, I use a trio of microcontrollers to make a set of servomotors appear to Lucy's brain like groups of biological muscles.

I'll tell you some more about these things when we look at other aspects of biological and mechanical bodies in the next few chapters, but while I'm on the subject of microcontrollers I want to share with you one of the blacker moments of my adventures with Lucy. Partly I just want to get it off my chest, but also I think it's important to show that, when exploring such deep waters, it's often the simplest things that scupper you.

Computers and I have a love–hate relationship: I love them and they hate me, or so it seems lately. Until now, I seem to have been blessed with a magic touch – computers and software that have been playing up and driving other people wild just roll over and beg to be tickled when I enter the room. Programmers who've been tearing their hair out for hours trying to find a bug are usually less than grateful when I idle past their workstation and the source of their problem catches my eye instantly. Building some of Lucy's systems using my favourite little PIC chips also turned out to be a breeze to begin with, and thanks to some lateral thinking here and there, the PICs were soon performing with a prowess that belied their minuscule stature. A couple of months into the project, all was well with the world.

And then I started to work on Lucy's hearing. I'd naïvely assumed that her eyes were going to be the most demanding problem, computationally speaking. A single frame from a video camera contains a lot of information, and the frames whiz past at a rate of knots. But it was, in fact, the mathematics involved in simulating the cochlea of the ear that proved too much for such a simple, slow, processor, and no matter what clever tricks I thought up to make it faster, it quickly became obvious that the PIC couldn't deliver. This meant changing to a new kind of chip, and that turned out to trigger a whole cascade of problems.

Electronics is going through rapid change at the moment. Chips used to be relatively simple, bulky devices, with big metal legs that could easily be pushed through holes drilled in a circuit board and soldered into place by a clumsy, ham-fisted person like me. The circuit boards themselves were therefore pretty easy to design and build too. But more modern chips come in what are called 'surface mounted' packages, with vast numbers of very, very tiny legs that need to be soldered to extremely fine copper tracks on the top of the circuit board, instead of being shoved through holes surrounded by copper pads the size of pennies. Surface-mount chips can be very densely packed, which is important for modern, highly complex consumer equipment. They are easy to place using robots, and to solder using expensive temperature-controlled ovens. But I'm not a robot, and the legs are so close together that one slip of the soldering iron, or one speck of dust on the photographic film when producing the circuit board, and the whole thing is ruined. Some of these chips can cost seventy pounds each, so mistakes can be traumatic.

Unfortunately, while the friendly little PIC chips are still available in the old through-hole form, all the more powerful processors with the right characteristics are of the surface-mount variety. So there was nothing else for it but to learn how to build surface-mount circuit boards. In fact it was worse than this, because a change of microcontroller brought up all sorts of other technical issues, which you really don't want to know about. Suffice it to say that I had a lot of new things to learn, as did my wife, Ann, whose more delicate and accurate fingers proved invaluable for soldering tiny components and etching the circuit boards, which she did with the help of her best lasagne dish and the kitchen stove.

What with this and other assorted knock-on effects, I ended up wasting about nine months. Nine months! You can usually get from conception to birth in that time. Unfortunately, when I started the project I only had about six months' money in the bank, so it's a good thing you bought this book – I need the royalties!

Aside from needing the money, one of my motivations for writing this book in the first place was to show that real science *can* be done by real people; that you don't necessarily need fancy laboratories and a PhD, just enthusiasm and an enquiring mind. I still think that's true, and I hope you'll agree with me as you read on, but I guess this stands as a warning that things can sometimes get a bit close to the edge.

Figure 7 Surface-mount chip, less suitable for clumsy fingers than the old-fashioned sort next to it

Being an independent scientist has its advantages over the academic route, and ordinary mortals like you and me are perfectly capable of making a real contribution to knowledge, or of making our own personal discoveries about the world instead of having them handed to us from on high by experts. But nobody funds people like me, and one never quite knows when the last straw is going to turn up.

Still, I'm soldiering on (or should that be soldering on?). Lucy now has a whole new set of 16-bit computer boards (see plate 3), this time connected up into a multi-processor array, with a biggish slab of shared memory that they can use to talk to each other. The computer boards are also conveniently identical, with little daughter-boards plugged into them to configure each one for a specific sensory device or motor system. I wish I'd been a bit bolder, because these computer chips are still not powerful enough for what I really want to do, and I've had to make some awkward compromises, but Lucy and I are at least back on track. And despite these inevitable relationship problems, my love affair with computers is still far from over.

CHAPTER SEVEN

All the better for seeing you with

The hero of every lone inventor has to be Philo Farnsworth, who dreamed up the idea of using a scanning electron beam to produce television pictures. According to legend, the idea occurred to him at the age of fourteen, as he moved back and forth on a horse-drawn harrow, tilling a potato field under the blaze of the summer sun. Farnsworth spent most of his life locked in a bitter struggle with the giant RCA Corporation over patent rights and hence, in the eyes of inventors everywhere (at least, those few who have ever heard of him), he is our patron saint.

My reason for mentioning Farnsworth is to point out the sheer brilliance of the concept behind television: turn a two-dimensional image into a one-dimensional string of voltages in a camera, squirt this along a single wire (or a single radio signal) and then scan it back again into a two-dimensional image at the other end. Two perfectly synchronised zigzag motions, one in the studio camera and the other in every television set in the land. Without this serialisation process, a television picture would need to be sent along more than a hundred thousand individual wires, one for each picture element (pixel), and to amplify or otherwise alter the signal we would need a hundred thousand electronic circuits. Yet this is exactly how the eyes and the visual circuits of the brain work. Essentially (and this will need qualifying shortly) every single pixel has its own 'wire' in the optic nerve and its own processing elements in the retina and the brain.

Unfortunately, thanks to Farnsworth, the video camera that Lucy sees with produces a standard television signal, which I have to turn back into a square array of binary numbers using one of her micro-controller chips before I can do anything useful with it. Once it is in this format, I can modify the data in such a way as to emulate the signal processing that occurs in the retina.

Ideally, of course, Lucy would have two eyes, but I decided I couldn't justify this to begin with, because depth perception requires reasonably sharp, high-resolution images and hence large numbers of neurons in Lucy's brain with which to process them. I simply don't have the computer power. So Lucy will have to manage with low-resolution vision from a single eye (a mere 32 by 32 pixels leaving her simulated retina, making 1024 signals in total). I'm already beginning to regret this, because some of the most interesting aspects of the visual cortex relate to binocular vision. But I must be practical, and since all these nerve signals have to travel down a wire from Lucy's on-board computers to the PC containing most of her brain, speed of transmission is a real issue. Something had to give.

Anyway, let's start at the beginning with a video signal, and look at the process the image goes through before reaching Lucy's brain. Lucy's eye is a tiny camera, about the size of a human eye, which produces a standard television signal. This is made up of several components, designed in such a way as to simplify the control of the electron beam inside an old-fashioned television set. Nowadays we would design things differently, but it isn't difficult to convert this standard analogue video signal into a two-dimensional array of numbers in a computer. A video frame (one complete image) starts with a characteristic pulse called the frame sync, which tells the electron beam to return to the top-left corner of the screen and begin its first scan. Then come a few blank lines to give the beam time to move into a visible area of the screen (these blank lines are nowadays used to transmit digital Teletext data). Thereafter, each scan line begins with a line sync pulse, which resets the electron beam to the left edge of the screen, followed by the colour data for that row of the image (or in Lucy's case greyscale data, since her camera is only black and white). The voltage of the signal rises and falls in proportion to the level of brightness at each point on that line of the image.

All of this happens very quickly. A complete scan of the screen occurs every fiftieth of a second (in the UK system), during which time about three hundred scan lines are delivered (British television has 625 lines, but these are delivered in two interleaved images of about 312 lines each, to reduce flicker). Therefore each line of the picture lasts about sixty-four millionths of a second. Since Lucy's initial (unprocessed)

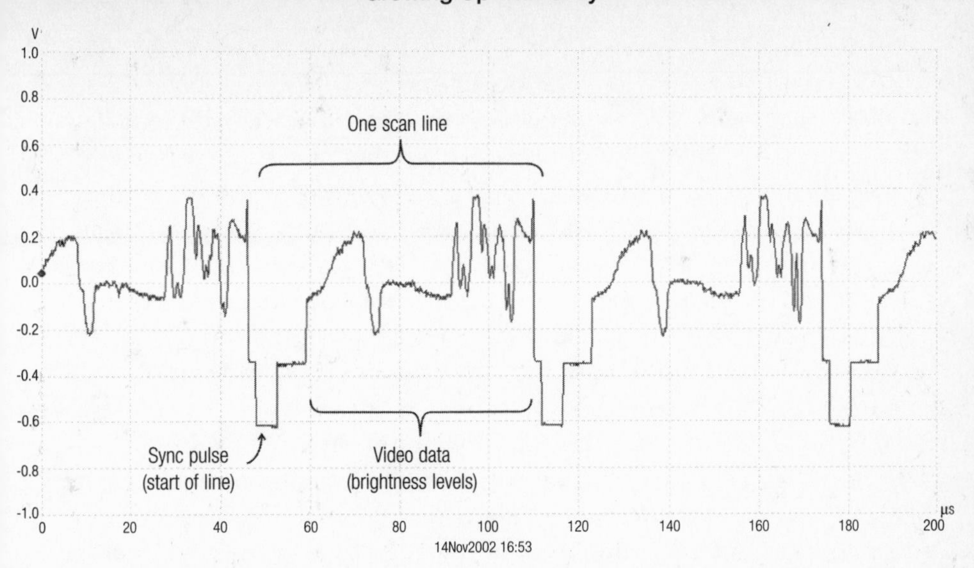

Figure 8 Oscilloscope trace of the video signal emanating from Lucy's eye

image resolution is 64 x 64 pixels, I need to be able to read the voltage on the video signal and convert it into a binary number about a million times a second. The little computer chips I'm using contain analogue-to-digital converter circuits (ADCs, see chapter 6) that can read voltages and turn them into numbers, but nothing like quickly enough. Luckily there are specialised ADC chips that can do this sort of thing with one hand tied behind their backs, so it is a relatively simple matter to use a microcontroller to control one of these chips and read the data quickly into memory. The MCUs I chose for Lucy are *just* capable of doing this quickly enough to keep up with the video signal.

The next stage is to make this image look more like the signal that travels down the optic nerve to the brain in real animals. Conventional approaches to computer vision do not do this. They assume that a fixed-position television camera, sampling absolute brightness levels, evenly distributed across the scene, is the right way to approach the problem. But in animals the visual image is very heavily modified during its passage to the brain, and these radical changes in data format are likely to be no accident, so for Lucy I need to take account of them.

One of these changes involves emulating the decidedly uneven distribution of light-sensitive cells in our own eyes. As you probably know, our visual acuity is extremely high in the central degree or so of our visual field (the portion of the image falling on a part of the retina called the fovea), and then drops off dramatically towards the edges. Obviously it never looks this way to us. The image entering our eyes seems to be crystal clear in every direction, but this is an illusion – the world looks detailed and sharp in every direction for the same reason that a cave looks brightly lit in every direction as you turn your head, if you happen to have a spotlight attached to your hat! If you hold your gaze steady and instead use the internal spotlight of your visual attention to look at objects in your peripheral vision you will notice that you can see little more than blobs of colour outside the middle few degrees of your visual field.

Now it could be that this is simply an evolutionary solution to a practical problem and Lucy doesn't need it. Getting all the nerve signals out of the back of a real eye without preventing the light from getting in, and without the optic nerve becoming so thick that it wouldn't bend properly, probably forced evolution to make compromises. On the other hand, this uneven visual acuity might have some genuine advantages. For one thing, if we have such a narrow 'beam' of sharp vision, we are obliged to move our eyes around the scene in order to see it all, and the very act of moving our eyes tells us something about what we are seeing. If we pan along a straight edge, then our eyes will move in a straight line, while the act of glancing around within the boundary of an object or surface may produce a characteristic set of eye movements that can tell our brains something about the object's shape. If useful information enters the brain, the brain will learn to use it, and records of eye movements are potentially a very compact and useful representation of the visual scene.

Vision is certainly an active process – we see because we look. In fact if we are prevented from looking, we quickly lose the ability to see. Normally our eyes are making small darting movements all the time (known as visual saccades). If the eye muscles are anaesthetised to prevent them from moving, or they are presented with a computer display that is made to move precisely with every microscopic eye movement and so remain static on our retinas, we become almost instantly blind. Perhaps these small saccadic movements are a way to turn static edges into moving features that are easier to pick out, but it is also possible that our brains keep track of the eye movements and use this information in addition to the image itself.

In a sense the visual world also acts as a kind of external memory.

We don't need to record the colour and brightness and shape of the entire scene in our brains, because to find these things out when we need to know them, we simply take a look. In this sense, the spotlight of our vision could be said to be 'addressing' this memory and accessing one small part of it at a time, as required, in order to answer our specific (unconscious) questions about what we can see. Such focused attention serves to minimise the amount of information we have to deal with at any one moment, and as we shall see later, makes perfect logical sense in relation to some other ideas I have about the brain.

There could be other ways in which our brains capitalise on this strange arrangement of sensory cells, or maybe it could suggest other illuminating questions we should ask. Take, for example, the way the optic signals are fanned out on to the surface of the brain. The signals from the optic nerve pass through various relay stations to reach the cerebral cortex, and then seem to be distributed more or less evenly across the surface of the primary visual region of cortex, known as V1. Because the distribution is even across the brain, but decidedly uneven across the retina, this means that the image becomes grossly distorted on the brain's surface. The middle degree or so of the visual field contains the vast majority of the sensory cells, and hence now occupies the largest area of the visual cortex, while the much larger periphery of our vision becomes reduced to a relatively thin ring around the edges. If you looked straight at me, the image of my face imprinted in the nerve activity on the surface of your cortex would magnify my nose to the exclusion of almost all else.

Does this have any significance? Well that depends partly on what V1 is actually doing with the visual information, which is a big subject that I shall come back to in the next section. The textbook answer is that V1 is detecting the orientation of the edges of surfaces, or at least enhancing them so that they stand out more, rather like the 'sharpen' tool in a photo-editing program. However, there is good evidence to suggest that this is only part of the truth, and V1 is also sensitive to oriented patterns with a range of given spacings (spatial frequencies). In other words V1 may be sensitive to textures as well as edges.

If so, then what should we conclude from the fact that these textures are heavily magnified on the surface of the brain if they lie in the middle of our vision, compared to how they look when they are off to one side? Are the pattern detectors larger in the middle of the cortical map, in order to compensate for this distortion? I can find no evidence

Figure 9 Image analysed by spatial orientation and frequency

to suggest that this is the case. If the range of spatial frequencies to which cells are tuned remains the same right across the visual cortex, then this means that the centre of our vision is sensitive to a finer range of grid spacings than the rest. So when a patterned surface moves across our visual field, why do we believe it to be the same surface, and not think that its pattern is changing size? Do we learn to correlate these detectors, so that we know which central detectors equate with which peripheral ones?

Alternatively, is it plausible that we make use of this shift in spatial frequency range to help us recognise objects regardless of distance? As a textured object gets closer, its texture gets coarser, but the object also covers more of our visual field and hence the larger texture falls on less 'magnified' portions of the brain and may trigger texture detectors with exactly the same physical spacing, allowing us to recognise the same texture from any distance using the same class of cell. This doesn't sound right, because the central part of the object will still appear to change scale, but it does show how tremendously important it is not to miss out strange aspects of the biological structure of sensory systems, just because they seem irrelevant. At the very least we have discovered an awkward fact that needs to be explained, and in explaining it we may gain insights that we otherwise would have missed.

Distorting Lucy's visual image to approximate that of the human eye is not difficult. The ideal way would be to fit a distorting lens on the front of her camera, but this would have to be specially made, and grinding my own lenses is one skill I think I'd be wise to avoid having to learn. Performing the distortion in software is less elegant but much easier to do, and this is why I started out with a 64 x 64 pixel image when 32 x 32 pixels will be enough for the final signal. By using the full image resolution in the middle portion, and then combining the remaining pixels into ever-blockier chunks towards the edges, I can produce a result that has at least a vague resemblance to the real thing. Looking 'through' Lucy's eye makes it pretty clear that she is never going to see very much with such a pathetic level of visual acuity, but it will have to do.

Another big difference between a photographic image and the signals that reach the brain through the optic nerve is this: in a digital photo, the numbers represent absolute brightness levels, but the retina is more interested in *changes* in brightness, either over time or across space. The absolute brightness of an object changes dramatically according to whether it is lit by the sun or hidden in shade, but the *pattern* of light and shade remains relatively consistent regardless of lighting conditions and hence is more use for helping us to recognise the object. The brain is particularly interested in edges, because these have characteristic shapes that enable us to recognise objects, and also help us to isolate objects from their background. To enhance the brain's ability to detect edges, nerve cells in the retina combine signals from many adjacent light detectors, comparing the intensity of light at the centre of a cluster with that at the edges. If the centre and surround are about equally illuminated, this part of the retina must be looking at a continuous surface; if there is a big difference between centre and surround, it must be looking at an edge.

What you *think* the scene looks like

What it really looks like to your eyes

How visual space is represented on the cortical surface

The cortex 'sees' contrast, rather than brightness levels

Figure 10 Impression of the 'distortions' caused by our visual system

If the centre is brightly lit, and the surround is dim, then subtracting the centre signal from the surround gives us a strong signal that tells us we are looking at a white line on a black background. Unfortunately, if we are looking at a black line on a white background, the centre will be dark and the surround will be bright. Subtracting the centre signal level from the surround would now give us a negative number, and since neurons can't represent signed numbers, the output from the cell would be zero – exactly the same as an evenly lit surface. So in real retinas there are two distinct populations of cells, one called on-centre/

off-surround and the other called off-centre/on-surround. One subtracts the centre value from the surround and the other does it the other way, so that between the two types of cells we can detect both black edges on a light background and white edges on a dark background, and distinguish these from continuous surfaces. So the visual signal reaching our brain is essentially made from two quite separate images, superimposed on each other: one measures how much darker each point is compared to its neighbours and the other how much brighter.

In practice the cells in the retina are even subtler and stranger than this. For example, some of them are more sensitive to movement than others, and individual cells can provide information about relative brightness, absolute brightness and changes in brightness all at the same time. But I can emulate this behaviour to a limited degree without much effort, and I can come back and do it more comprehensively if it turns out to be helpful. There are other clever tricks that should be borne in mind too, like the fact that some nerve fibres leaving the eye conduct signals more quickly than others. One way in which the brain might capitalise on this is to detect movement, because the signals in the fast and slow pathways will differ when an edge is moving across them.

If all this wasn't enough, here's another strange fact: most of this preprocessing of the visual image occurs in the retina itself, rather than in the brain, particularly in large nerve cells called ganglion cells. Each of these ganglion cells gathers together inputs from many primary light-sensitive cells, partly to support this centre/surround functionality and partly in order to amplify the extremely weak signals produced by photons hitting the individual cells of the retina. Because of this signal-gathering effect, the visual image gets spread out and blurred quite a lot at this early stage, so that a single point of light will stimulate many optic nerve fibres, and a single nerve fibre will respond to quite a wide region of visual space. Guess what happens when the signals get to the brain. Do they get focused back into finely resolved points again? No, they seem to get spread out and blurred even more. And more, and more, as the signals travel along the various cortical pathways. What arrives along the optic nerve is a deeply distorted, non-greyscale, blurred image, and then the brain makes it worse. So how is it that we can tell how bright something is?[1] More to the point,

[1] To some extent we can't, and there are various optical illusions that demonstrate our inability to make accurate judgements of brightness, but we can still do better

how can we see any fine detail at all, if the image is so blurred? In fact we are able to see details that are even finer than the theoretical resolution of our eyes. Clearly, something odd is going on here.

At the very least I think we can state a useful rule about theories of vision. Whatever abstraction and computation we think might occur during the process that turns a bunch of dots of light and shade into a perception of objects located in three-dimensional space, *no information can be lost in the process*. If the image gets blurred through being repeatedly 'gathered together' by nerve cells (convergence) then it must be done in such a way that the original image could be reconstructed. It doesn't mean the original image actually does get reconstructed anywhere – that would be pointless, since there is no little man inside our heads to look at the image and interpret it (and his own visual system would blur it again anyway). But if someone asks me to read a signpost that is on the limit of my retina's acuity, I can do it, yet the image in my brain is almost as blurred as if I was looking through frosted glass. If they ask me what colour something is or whether one thing is brighter than another, I can tell them. I can recognise the letter 'F' regardless of which way up it is or how distant it is, as if these factors had been completely removed from the equation, and yet I can still *tell* you which way up it is and how far away it is. This is not magic, it's mathematics – a blurred image that has been blurred in a known way can be reconstructed. This is how the blurring caused by a faulty mirror in the Hubble Space Telescope was removed. But it certainly makes the problem harder to think about, and it suggests that the way the brain processes information might be rather counter-intuitive.

than one might imagine, given that the nerve signals are not measures of brightness but measures of spatial changes in brightness.

CHAPTER EIGHT

Why the best cocktail parties need superior olives

Speaking of blurred images, imagine you are at a cocktail party. Everybody has had a few martinis, and by now most people are talking rather too loudly and laughing rather too boisterously. Suddenly, out of the corner of your ear, you hear someone mention your name. You carry on nodding wisely at the things your host is saying to you, but pay little attention to the actual words, because your ears are searching around the room for the one conversation that sounds like it might be interesting, since it's about you.

This capacity to identify and single out one sound among many is called the Cocktail Party Effect, and is just one of the amazing things that our auditory system can do that we never really think much about. How does the brain do it? I'm afraid I don't know. However, when I started work on Lucy's hearing, with the help of a summer student, Tom Groves, we developed a mechanism that could conceivably have some bearing on the matter. Ostensibly, I'm telling you about this because it shows that thinking about biology from the point of view of a practical engineer can lead to useful results, even in the absence of specialist knowledge. In practice, though, I'm telling you because the part of the brain we were modelling when we found these results is called the superior olive, and the irony was too good to miss.

The superior olive is one of the lower, more primitive parts of the brain – one of the highly specialised bits that don't really interest me much. Even so, I still need to model them because these are the kinds of structure that communicate with the cerebral cortex. Nerve impulses from the two ears arrive at the superior olive and are combined with each other before passing on to the cortex and to a part of the brain called the inferior colliculus. The olivary complex is known to be associated with our ability to work out which direction a sound is coming from. For high frequency sounds we achieve this simply by measuring the sound's relative loudness in each ear, but for low frequency sounds we instead detect the phase relationship. In other words, we measure the minute delay between the times at which the same part of a sound wave reaches each ear. If both ears hear the rising edge of the waveform at exactly the same moment, the sound must be directly in front of us or behind us.[1] If it hits the left ear before the right, then the sound is to our left.

One theory about how this phase relationship is measured is that the signals entering the two ears pass in opposite directions along a chain of neurons that act as a delay line. The signal from the left ear enters the left end of the chain and propagates towards the right, while the signal from the right ear enters from the right and propagates left. As the two signals move past each other, they are somehow compared to produce a measure of their phase difference. So Tom and I decided to test this idea out and give Lucy the ability to detect a sound's direction.

Imagine a sudden handclap directly in front of Lucy's face. The sound reaches each ear-drum (a microphone) at the same moment, and travels in opposite directions along a virtual delay line, programmed into her auditory microcontroller. The two peaks will coincide when they reach the middle cell of the chain, and from this we can tell that the sound is central. If they coincide towards the left edge of the chain, we would know that the left-going signal had been travelling along the chain for longer than the right-going one, and hence must have reached the ear-drum first, meaning the source of the sound must be to the right. With continuous sounds like speech it is a little more complex than this, but essentially all you have to do is subtract the left-going signal from the right-going one and look for the point where the signal is weakest (where the two waveforms best cancel each other out).

Alternatively, if you add the two signals together, the point of best

[1] We seem to be able to differentiate between in front and behind partly by listening to the difference in timbre caused by the sound passing the pinna (ear-flap). However, it is also possible that small head movements can pick out the difference, because as we move our head to the left, a sound from in front of us will appear to move to the right, while a sound from behind will appear to move towards our left ear.

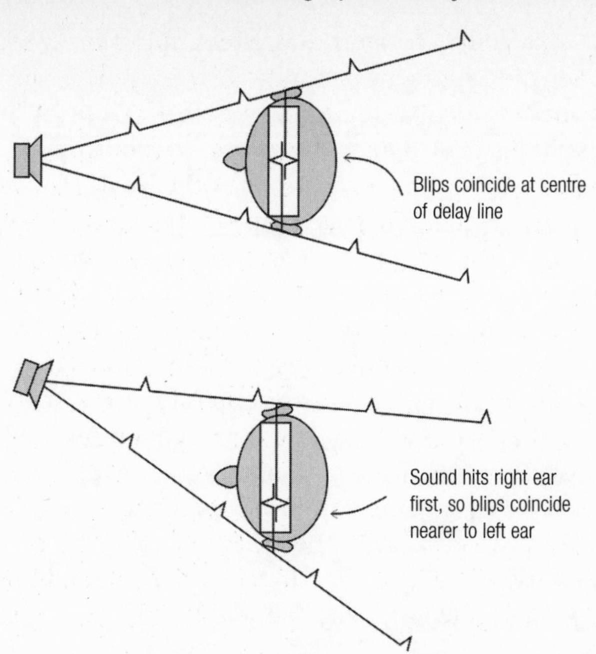

Figure 11 How the superior olive may detect the direction of a sound

fit will produce the largest signal on average, and if there are two simultaneous sources of sound you will get two peaks. By listening to the signal produced at one selected point on the chain of neurons, you can pick out a single sound in relative isolation from all the others, since sounds coming from that direction will be maximally reinforced. Hence the cocktail party effect may partly depend on having superior olives! It is perfectly possible to tap off at various points along the chain to create a 'map' of the locations of all the sound sources, relative to the direction of the head. Such a map is exactly what the inferior colliculus needs in order to control the head movements that unconsciously cause us to turn to face an unexpected sound.

Whether our conscious internal attention mechanism can and does manipulate this map in order to tune into a single conversation or pick out the rustle of a predator from among other noises I don't know, but all the brain would need to do is selectively enhance the signals in one region of the map. If it enhances them a little, we will be able to pick out one particular sound more clearly from its background; if it enhances them a lot, it will trigger a reflex head and eye movement that causes us to turn towards the source of sound (and feels to us like a conscious decision to move). There are strong parallels between this ability to attend deliberately to a sound and our facility to attend either internally or explicitly to a portion of the visual scene, both of which involve the same set of brain structures.

I offer this as an example of how, if you think about the problem from a biologically valid perspective, starting with little more than a blank sheet of paper, you can devise a solution that may well match what happens in real animals. Notice the interesting relationship between attention and intention; notice the way an involuntary reflex has been turned into a voluntary action; notice also that we use neurons as computing elements here in a very different way from the way that they are used in most neural networks (see chapter 3).

What else do we need to supply if our future androids are to be able to learn to hear? One of the key non-brain computations carried out by our auditory system is the complex frequency analysis performed by the cochlea. As I'm sure you remember from all the tedious diagrams you drew in school biology, the sound waves hitting the ear vibrate the ear-drum, which then wiggles a group of tiny bones that transfer the amplified vibration to the oval window of the cochlea. The cochlea itself is a chamber shaped like a snail shell, filled with a liquid that resonates at different points according to the diameter of the tube and the frequency components of the sound. These resonances are then picked up by tiny hairs, which trigger nerve impulses. For high frequency sound (low frequencies are encoded differently) any given neuron will carry a signal that represents the loudness of one specific frequency. All natural sounds are composed of a wide variety of frequencies at once (only a pure sine wave has a single frequency), and it is the relative loudness of each frequency band that defines the characteristic timbre of the sound. By the time the signals reach the brain, these complex intertwined waveforms have been separated out into a spatial pattern, with high frequencies at one end and low frequencies at the other. Such a layout is called a tonotopic map, and every type of sound will produce a distinctive spatial and temporal pattern on that map.

It's interesting to note that the cochlea essentially paints a picture with sound. This corresponds closely with the literal picture presented to the brain from the eyes, and also to the various pictures of the body that are painted on the somatosensory (touch and joint position) and motor (muscle tension) maps of the cortex. Most or all of the primary cortical maps therefore work with pictures and I think we can presume that even the higher cortical functions such as plans and words and goals are represented in a similarly 'pictorial' way. More on this later.

Much as I would have loved to carve Lucy a pair of synthetic cochleas from plastic, and placed thousands of microscopic vibration sensors inside them, my fingers simply aren't small enough, so Lucy's cochlea has to be virtual. There is a fancy mathematical technique for breaking complex waveforms (whether they represent sounds, earthquakes, radar echoes or stock market fluctuations) into their constituent frequencies, known as the Fast Fourier Transform. Unfortunately, because of my various disasters with finding suitably powerful micro-controllers, I didn't have enough computer power available to perform a Fourier Transform at the same time as all the other things Lucy's auditory circuitry had to do. So instead I wrote a computer model of a series of springy weights of different sizes that would resonate at characteristic frequencies, and hence would be vibrated by a sound signal in proportion to how much of their characteristic frequency was present in the sound. To those who know about Fourier Transforms, this probably sounds even harder, but in practice it is very quick to compute because it doesn't require any multiplications. It also feels much closer to the spirit of the biology.

While I'm on the subject of sound, and since I'm itching to get on with section three, which is about how the brain works, I'll talk about Lucy's voice now. You might think that the sensible thing to do would be to take advantage of all the developments in speech technology and give Lucy a standard speech synthesiser chip, so that she can speak in plain English. You might think that, but you'd be wrong. If I were trying to jump to the moon then this would be a good approach. People would be very impressed if Lucy could speak real words, and it would look like she was really intelligent. But what would she say?

Brains do not deal in words and sentences (let alone ASCII computer codes); they deal in the language of muscle movements. It may seem to us that we formulate sentences in our heads and send them off to some low-level mechanism that worries about converting the symbolic words into muscle sequences that drive the throat and lips. But how do we know that this is what really happens? How often do you actually know what you are going to say before you hear yourself say it? Do your utterances really start out as complete grammatical sentences made from words, which are in turn constructed from letters or phonemes? Maybe they do, or maybe they remain stored in the form of a hierarchical dance of muscles and never become sequences of symbols unless, and until, we write them down. We have almost no idea how language is represented in the cerebral cortex. All we know is that the end result is a set of muscle movements, and that we learn to correlate these movements with the sounds that they make through hearing ourselves speak.

Perverse as it may seem, therefore, I decided that the best way to use speech in order to help me understand the brain was to copy nature as closely as I could and saddle Lucy with the task of working out how to manage her own breathing, vocal cords and mouth movements. In practice, all she has yet managed to produce is a series of vaguely vulgar burps.

It seems to me that the way babies babble and then listen to what comes out of their mouths may be crucial for the formation of the patterns in their brain that they later build on when learning to speak. Understanding what someone else is saying may depend in part on our ability to learn what mouth movements we would have to make ourselves to form the same sonic pattern in our auditory cortex. Associating a pattern of muscle movements with a pattern of sound frequencies, and vice versa, is an interesting and probably very illuminating challenge, which I would miss completely if I gave Lucy a ready-made speech synthesiser.

Again, the trick was to use a computer to build a virtual machine – an analogue of a real vocal tract that exists only in the cyberspace of a computer's memory. It is very easy to simulate the way that the diaphragm responds to nerve impulses to draw in and expel air. It is also easy to implement a mathematical model of the glottis, tongue and other assorted flaps of skin that we use to hold in a breath or to break up an out-breath into a series of explosions (say 'pat-a-cake' a few times and you will feel which flaps of skin I'm talking about). Producing a sound from Lucy's virtual vocal cords is also quite straightforward, since a sound waveform is simply a repeating sequence of numbers representing the height of the wave at any moment. Sending these numbers to a digital-to-analogue converter and thence to a loudspeaker will produce a sound at a pitch determined by how quickly the sequence repeats and at a timbre dictated by the shape of the wave.

The hard part is simulating the way the mouth and other cavities behave as resonant structures. When we say 'ay, ee, eye, oh, you' we achieve the different vowel sounds by changing the shape of our mouth, which forms a resonant cavity that accentuates some frequencies in the sound while suppressing others. The relationship between these resonances (called formants) and the resultant sound waveform is not straightforward, but various mathematical techniques exist to make the computations possible, and since this class of mathematics has a possible bearing on some of the things that go on inside our brains when we think, this is a good opportunity to illustrate some of the ideas.

One of the most useful building blocks in electronics is the filter. The bass and treble controls in a stereo are filters. The bass control is called a low-pass filter, because it removes all the high frequency components from a sound. The treble control is called a high-pass filter for equally obvious reasons. When you hear a (proper analogue) music synthesiser make that delicious 'eeeeeeeeuuuuuoooooowwwww' sound, you are hearing a band-pass filter sweeping down the frequency range, selectively boosting a narrow band of frequencies in an otherwise constant sound source. The various resonant cavities of the vocal tract are essentially band-pass filters, and the difference between the sound 'oh' and the sound 'ee' is due to the resonant frequency of that filter. One way to simulate filters in a computer is to use the mathematics of the Fast Fourier Transform to separate the sound into its constituent frequencies and then modify their relative strengths before mixing them back together again. Another way (and the way I decided to use in Lucy) is to program something called a Finite Impulse Response filter. These are surprisingly easy to understand with an example.

Suppose you and I are standing at opposite ends of a tunnel. When I make a continuous sound with my voice, you will hear a much more booming rendition of it at your end of the tunnel, because the tunnel preferentially resonates at certain frequencies, and thus enhances these frequency components in the sound. Now suppose that I make a sharp 'click' noise. What you will hear at the other end is a somewhat drawn out sound, consisting of the initial click and the subsequent fading echoes from the tunnel walls. These echoes are produced by the walls of the tunnel through precisely the same process that causes them to resonate when the sound is continuous. What is more, a continuously varying sound wave is essentially a long sequence of clicks, following one immediately after the other at different loudness levels. If one click produces a characteristic, decaying echo, then it follows that a series of adjacent clicks will produce a series of overlapping echoes. As the sound wave changes in amplitude, so will the overlapping echoes. So, to simulate the effects of a tunnel on a continuous sound wave, we simply have to know what the characteristic pattern of echoes is for that tunnel. This is known as its impulse response, since it describes how it responds to a short impulse, or click. We can then break the initial sound wave up into tiny chunks (impulses), and replace each impulse with a complete echo waveform, at a volume proportional to the height of the sound wave at that point. The sum of all these overlapping echoes will give us the waveform of the sound after it has passed through the tunnel.

This is a beautifully elegant technique, which you will find used in all sorts of digital signal processing applications, from mobile telephones, to CD players, to submarine sonar. More of your everyday life depends on finite impulse response filters than you might ever have guessed. The business of splitting a signal up into discrete impulses and replacing each impulse with a complete waveform of the same signal strength is just part of a much larger concept called convolution. In the previous chapter I talked about how the ganglion cells in the retina (and the cortical cells in the brain) gather together many pixels to perform centre/surround calculations for enhancing edges. You can think of one ganglion cell drawing convergent input from many adjacent light-detecting cells or, because of the way the cells overlap, you can think of each light-detecting cell as spreading its output among many ganglion cells. This process is reminiscent of the way that a series of discrete clicks become spread out into a series of overlapping echoes in a finite impulse response filter, and indeed both processes are examples of convolution. Convolution and filtration are very closely allied concepts, and when the brain fans out its signals (as it does almost everywhere), it is often reasonable to describe it as filtering the information and hence either removing something from the signal or measuring the amount of that something.

Convolution (and its converse, deconvolution) may be a crucial and universal concept in understanding brain function, but I don't want to get bogged down with mathematics in this book (or ever, for that matter). So just bear convolution in mind as you read the next section, and keep an eye out for examples of this kind of blurring, spreading-

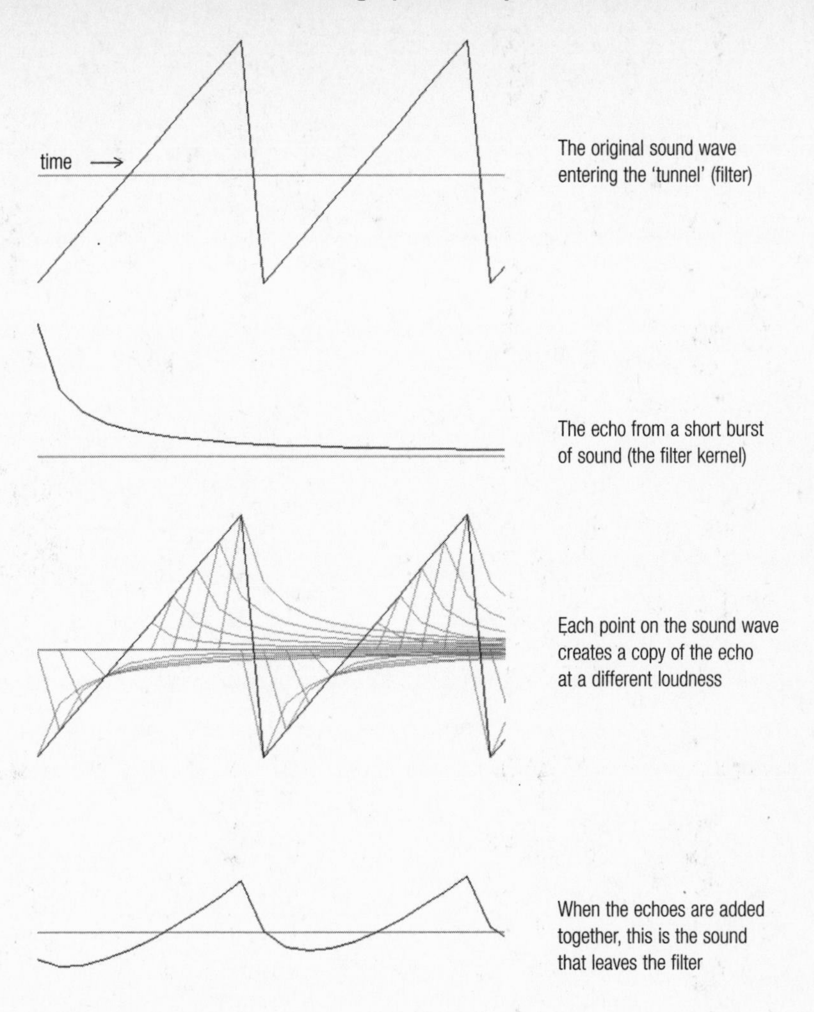

Figure 12 How a finite impulse response filter works

out process in action. It turns up in many different guises and plays many important roles.

I knew some of these computational ideas before I started work on Lucy; I learned a few more from Tom, our summer student. But only when I came to do the actual work on Lucy's hearing and voice did I realise quite how universal these concepts are, and how much they help me to think about the brain in general. If I hadn't built Lucy's body myself; if I had farmed the problem out to a team of people, as is standard practice in academic circles, I would have missed out on an awful lot of insights. Aside from being a control freak, this is one of the main reasons I decided to take on this project alone. Teamwork is a wonderful thing in its place, but breaking big problems down into smaller ones can sometimes be a big mistake. The only real snag about not being part of a huge team is that I don't get invited to so many cocktail parties.

PART THREE

Mind

Now comes the tricky bit.

It's at this point, when all the hardware has finally been built, that the thoughts of robotics enthusiasts turn to the question of software. For some, this is a bridge they planned not to cross until they came to it, and now find the chasm is rather deeper than they had anticipated. For others it is something to be quite pragmatic about – the end is more important than the means. Even for professional scientists, who wouldn't be building a robot at all unless they had some specific theory to test, the bulk of the problems associated with getting the robot to behave sensibly usually lie outside their area of research and are more of an irritation than anything else.

In the majority of cases, therefore, the general plan is to link together bits of existing theory, maybe adding a new twist here and there, to produce solutions for each of the individual problems facing the robot. Vision, for example, might be tackled by combining So-and-so's theorem with the approach invented by Such-and-such. The Whatsit technique might be extended to deal with the problems of navigation, and Thingummy's algorithm would be pressed into service for handling speech.

For Lucy, this approach would miss the point entirely. My aim with this project is not to extend existing methods. As I tried to explain in Part One, after twenty years of thinking about these things I've come to the conclusion that none of the existing approaches is going in the right direction. Nor is my aim to assemble a pot-pourri of conceptually unrelated mechanisms in the hope that the result will be intelligent. Seeing, hearing, speaking and moving are not different problems, in my mind, they are different viewpoints on to the *same* problem, namely: how does nature create general intelligence?

And this is the bridge I now have to cross.

CHAPTER NINE

Your brain: a quick-start guide

I recently vowed that whenever a journalist asks me questions about my project, I will ask them in return how *they* think the human brain works. What do they think is going on up there? What kinds of images and metaphors do they have for how their own minds and brains function? Doesn't everybody think about this from time to time, and develop some kind of theory about what they are and how they work? Most journalists are pretty smart people; many are very highly educated and thoughtful, so I've been expecting a lot of interesting insights. Unfortunately, so far I haven't managed to elicit much more than a blank stare. At best they show initial enthusiasm for the question, launch into a sentence that doesn't quite go anywhere, then tail off. Before I know it they've cleverly turned the conversation around somehow and I find that I'm back to fielding their questions, while mine remains unanswered. Presumably this is why they are journalists and I am not.

So what do *you* think goes on inside the lump of pink porridge that is your brain? Do you know your way around its various structures? This is the biggest problem when writing science books: the author has no chance to find out what the reader already knows, and therefore can't be sure how to pitch the new material, or which misconceptions need to be addressed. You could be an expert neuroscientist or you might have only the vaguest idea what a brain actually looks like; I can't tell. The only thing I can be sure of is that you are interested in the subject – after all, you've inherited a powerful piece of technology inside your head but it came without a user guide. This is not a book about the brain *per se*; there are plenty of those around already. My aim is to document the first steps in my attempt to make an artificial being, but since I'm using natural living things to provide most of my clues about how to do it, and the brain is such a puzzling object at almost every level, I guess I'd better lay out a broad map of the terrain quite early on. Let's grab a mirror and a really sharp hack-saw, get the hood opened up and see what we can find.

The first thing to notice, once you've removed your skull bones, is that you appear to have a cauliflower growing in your head (see plate 4). Untold numbers of long nerve fibres enter your brain along the spinal cord and as many again project down it in the other direction, ending up in the various muscles, glands and tissues of your body. Still others arrive and depart through the cranial nerves of your head (I learned their names in school – I recall the mnemonic perfectly, but for the life of me can't remember what any of the initials stood for or why on earth anyone thought it was worth knowing). Almost all of these fibres meet in the deep, central core of the brain (the brainstem) from which billions of secondary fibres fan outwards and upwards, as if your spinal cord had become badly frayed, and the brainstem was a knot someone had tied in it to stop things getting any worse.

Actually, the cauliflower analogy is better than that of frayed string, since string has a fixed number of fibres, whereas the brain, just like a cauliflower floret, has far more fibres in its outer regions than it does in the middle. This is partly because the signals entering from the spinal cord branch at various points and partly because the outer layers of the fibrous 'white matter' of the brain also contain many loops that carry internal traffic from one part to another. The splayed fibres are interrupted here and there by the grey masses of various sub-cortical structures and then finally erupt at the thin layer of nerve cells forming the skin of the cauliflower, known as the 'grey matter' or cerebral cortex. The sub-cortical lumps and the thin grey rind of the cortex contain both the cell bodies of the neurons, whose long fibres make up the white matter, and unbelievably vast numbers of short-range interconnections.

The next thing to notice is that you have two distinct halves to your brain, as if someone had driven a meat cleaver through it when you weren't looking. By an accident of development, the signals from sensors and motor systems on one side of your body cross over the midline and enter the opposite hemisphere of your brain. Vast tracts of nerve fibres, the largest of which is the corpus callosum, join the two hemispheres together and make these unilateral signals available to both sides. In certain cases of severe epilepsy, the corpus callosum is surgically severed, to reduce the spread of the epileptic seizure and make life more tolerable for the patient. This shows that the two halves of the brain are capable of working more or less independently, even though these two separate brains are obliged to share a body and have rather different specialisms and interests. This leads to some very odd behaviour from the patients and gives a whole new meaning to the phrase 'being of two minds about something.'

Already the brain is starting to look like a very odd machine indeed. You can take a knife and slice it in two, and it still works – try doing that to your PC and one fundamental difference between a brain and a computer will quickly become obvious. Another equally grotesque surgical operation, thankfully no longer performed, is the prefrontal lobotomy. In one common form of the procedure, an ice-pick (honestly) would be inserted into a poor psychotic patient's eye socket with the help of a good blow from a hammer, then waggled around until it had severed the front part of their brain from the rest. Anybody would think this sort of behaviour would *lead* to psychosis, rather than cure it – I'm pretty certain I'd be livid. Nevertheless, it sometimes had a calming effect on the patient and hence, during the nineteen-forties, became an extremely popular operation for a wide variety of illnesses, including such obviously incapacitating psychiatric disorders as 'being of the wrong political persuasion' or 'being an irritating relative'. In fact, lobotomy only improved about a third of the patients, while another third showed no change and the remainder got worse, which is about the same proportion you would expect if you simply left them to their own devices and didn't insist on poking them in the eye with a sharp object. However, the interesting point is that it is possible to take an ice-pick and a hammer and screw around with someone's brain with little or no discernible effect. Rather than a delicately balanced, incredibly complicated and detailed mechanism, it's enough to make one think that the brain is merely there to stop the skull from rattling.

Nevertheless, in some parts of the brain quite small amounts of damage can lead to severe disability. In the case of the patient famously known to psychologists as 'HM', an operation similar to that of a prefrontal lobotomy was performed on his temporal lobes, with the effect that he was henceforth completely unable to lay down new memories. This leads us to another observation about the brain: different bits of it do different things. The part that was severed in HM is called the hippocampus, and this structure is therefore implicated in the process of memory formation. Other parts also suffer highly characteristic losses and syndromes when they are damaged or surgically altered, and this can often be revealing. In fact these 'lesion studies' are one of the main ways of studying the brain. For example, damage (usually through stroke or trauma) to a part of the occipital lobes can sometimes lead to the strange condition known as 'blindsight', in which the patient claims to be completely blind, and yet is able to perform visually guided tasks, such as inserting a letter into a slot. This shows that different aspects of vision are handled separately; a conclusion backed up in more detail by other lesions that, for example, destroy the ability to visualise colour or recognise objects. Sometimes this damage is highly specific, involving the recognition of familiar faces, or even the ability to discriminate between different types of vegetable.

You may feel that this localisation of function is rather obvious, but there are some complications. The first of these is that we know that one part of the brain or another is *involved* in certain aspects of behaviour, but we can't conclude from this that it *performs* these processes. Making this mistake is like deciding that a telephone makes pizza, because that's the object you speak into when you want one, or that a motor car is powered by water because it eventually stops working if you forget to fill the radiator.

The second misleading thing about the relationship between structure and function is the intuition that nerve activity implies computation. This is a real problem when using brain scanners to map out the brain. If a particular part of the brain becomes more active whenever someone performs a certain kind of task (such as using language or performing mental arithmetic) then it is tempting to conclude that this part of the brain is specialised for performing that task. But things are by no means this straightforward. Suppose neurons in region A send inhibitory signals to a distant region B. Is the computation occurring at A or B? Is the activity seen at region B really due to the increased activity at the tips of region A's output fibres, which are actually inhibiting region B, rather than exciting it, suggesting that the apparently active region is actually being turned off, rather than on? Suppose (and this is very relevant to future chapters) that we are really witnessing a dynamic balance between two opposing forces: one excitatory and one inhibitory. In this case all sorts of difficulties arise, because the total amount of neural activity tells us nothing at all about the balance between excitation and inhibition. The region could swing from full excitation + zero inhibition, through half excitation + half inhibition, to zero excitation + full inhibition, yet the sum of excitatory and inhibitory activity (and hence the strength of the scanner signal) would remain constant throughout. Or the total activity might change without the balance being affected.

In the worst case, the entire cortex of the brain might be acting as a single unit, so that every new source of signals alters the pattern of activity right across the brain. This is rather like the way echoes travel around a labyrinth of tunnels, so that the direction of the sound doesn't necessarily equate with its origin. Given that the brain scanner experts massage their figures to remove a great deal of the (supposedly irrelevant) activity before publishing their results, some of these problems may make the whole brain mapping enterprise extremely misleading, and it would be dangerous to jump to conclusions about the localisation of function from scanner evidence alone.

Nevertheless, the cerebral cortex is pretty clearly divided up into a series of patches or maps, each with a characteristic microscopic structure and a different function. The differences between maps are subtle, and the maps themselves are labile variations on a basic theme, rather than distinct structures. Cortex is cortex is cortex, but different cortical maps have slightly different proportions of cell types and slightly different patterns of wiring. Many of these cortical maps were detected using microscopes and tissue stains by Korbinian Brodmann, at the beginning of the twentieth century. Brodmann gave each map a number (in the order that he discovered it), so the map that receives the initial visual input from the optic nerve (via the thalamus – of which you'll hear more later in this chapter) is known either as V1 (for visual area 1) or Brodmann's area 17.

The maps are laid out on the convoluted surface of the cortex rather like the patches on a random and badly ruckled patchwork quilt. If we slice through the cortex and view the cut edge from the side, we find that the cells in each map are arranged in a series of layers. By convention these layers are numbered from I to VI, starting with the outermost layer. In many maps one or another layer may be almost non-existent, or it may be overdeveloped and subdivided. Primary motor cortex (M1, or Brodmann area 4) for example, is the primary source of outputs from the cortex to the body's muscles, and is characterised by having only a very faint layer IV. Primary visual cortex

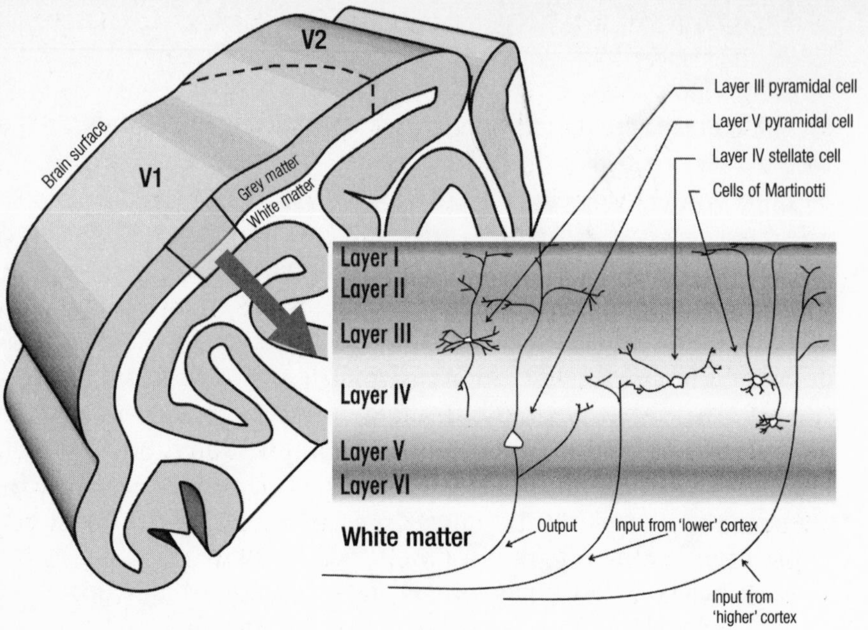

Figure 13 Section through the cerebral cortex, showing the major types of neuron

(V1), on the other hand, has a highly developed layer IV, subdivided into noticeable sub-layers, with a horizontal structure that looks like a pattern of superimposed bands and blobs, as if this particular patch of the quilt had been made from layers of stencilled fabric.

Generally speaking, though, each map has six layers. This laminar structure can be considered the archetypal structure of cortex, except near the curled-under 'fringes' of the cortical sheet, where the six-layered neocortex gives way to a three-layered structure called the hippocampal system, which is evolutionarily older and more specialised. The six layers are a bit misleading, in fact, since some are marked by the presence of certain cell types, while others are detectable largely because they contain strong horizontal bands of fibres. As a rough guide, layer I (the outermost layer) has few neuron cell bodies and is mostly made up of horizontal axons and dendrites rising from the layers beneath. Layers II and III contain many smallish pyramid-shaped cells and the junction between the two layers is often indistinct. Layer IV is marked by the existence of several types of small star-shaped (stellate) cells and layers V and VI contain large pyramidal cells, as well as other classes of cell. Strong bands of horizontal fibres exist in several of the layers, carrying signals locally within a single map.

As to the various cell types, the bulk of cortex is composed of pyramidal cells. These have long axons, often protruding out of the bottom of the map into the white matter, where they continue on up to other cortical maps or descend to sub-cortical structures. Pyramidal cells also have two main input sites, one on the cell body itself (basal dendrites) and one long tendril called the apical dendrite, that projects upward through cortex. Other common cells are stellate cells, which tend to cluster around pyramidal cells and provide them with inputs from a fairly wide radius. Some stellate cells are excitatory and some are inhibitory, and it is often the case that any given pyramidal cell will receive excitation from cells within a narrow radius, and inhibition from a larger radius (the on-centre/off-surround arrangement that we discussed in chapter 7). In addition, there are a range of other cell types (such as the cells of Martinotti), some of which point in the opposite direction to the pyramidal cells, feeding back their outputs towards higher layers and making the cortical sandwich highly recurrent (see chapter 3).

On average, each cortical cell has around ten thousand connections to other cells, with most of these connections being local, so in this respect cortical tissue can be thought of as having an architecture somewhere between that of a telephone exchange, in which all the signals are one-to-one, and a public address system, where everyone hears a bit of everything that is happening in the local area.

Given perhaps a million cells per square millimetre, and ten thousand interconnections per cell, it's no wonder that we don't know exactly how all this is wired up. Frankly, it's a bit of a mess in there. Neurons are remarkably gregarious objects that will grow a connection to anything that emits a signal, so it is difficult to know which connections are functional and which are just brief acquaintances or random accidents. Only the vaguest outline of the wiring is known, which of course is great news if you want to champion your own theory of brain function, because you can find evidence for pretty much any circuitry you like. But it makes deductions and generalisations extremely difficult and there are ample opportunities for any simplifications to be misleading. Nevertheless, if Lucy is going to get ahead, she has to get a brain, and to make conceptual leaps that might take me somewhere I had to try and form a generalised model for how the various maps are wired together – which layer on one map sends signals to which layer on another. As we will see in chapter 12 this generalised model leads to some interesting and potentially valuable conclusions.

For the moment, though, just treat each cortical map as a distinct processing module that receives parallel inputs from the senses (via the thalamus) and/or from other cortical maps, and sends outputs to the body (via other sub-cortical structures) and/or to other maps. The map-to-map connections are quite complex and can fan in and fan out. That is to say, several maps can send signals to a common destination map, and one source map can send signals to several destinations. There are currently thought to be around forty maps in the primate visual system alone, and any randomly selected pair of these visual maps is more likely to be interconnected than not.

Down deep in the brain, within or beneath the white matter, are other structures, many of which communicate with the cortex. As a general principle, the brain seems to have evolved through several phases during the evolution of vertebrates and each new phase builds upon the preceding ones. At the most primitive level, down in the brainstem, we have a network of neural modules that have been hard-wired by evolution to handle highly repetitive responses, such as the maintenance of the heartbeat, control of the digestive system or simple behavioural reflexes. On top of this, evolution has developed new structures, culminating with the cerebral cortex, which modulate or superimpose their will on these lower systems, often by suppressing or triggering innate reflexes. Lucy needs many of these subsystems, but to describe them here in detail would be tedious for both of us. I shall just mention three important structures that will crop up from time to time in the following chapters.

The first of these is the thalamus. All sensory and motor signals pass through the thalamus on their way to and from the cortex. So do many of the inter-cortical signals that pass from map to map. As a general rule, everything is reciprocally connected, so that a signal from the senses to the cortex via the thalamus will also give rise to signals from the cortex back to the thalamus. In fact, the bulk of the signals passing between cortex and thalamus travel in the opposite direction to the sensory data – most signals in the brain go the 'wrong' way, according to conventional wisdom. The thalamus is therefore a nexus for all the major feedback loops between cortical maps, and a place in which all these feedback loops can be modulated (and can modulate each other).

Another set of notable structures is the basal ganglia. The only thing worth saying here about these clumps of sub-cortical neurons is that they seem to be involved in gating and triggering muscle responses. Clearly one cannot move one's leg forwards and backwards at the same time. Consequently, a decision to take one action may render other actions impossible or undesirable. On the other hand, it is perfectly possible, up to a point, to use your left hand for one job while your right is doing something else, so not all decisions are mutually exclusive. Dealing with this exclusivity and prioritising one action over another seem to be the main roles of the basal ganglia. One thing we can infer from this is that at the cortical level, many actions might be vying for attention simultaneously – the cortex is not necessarily coming to a single conclusion, and many parts of cortex may be working independently of each other. The basal ganglia perhaps help to arbitrate over all these parallel decisions.

Finally, I should mention another group of structures, the colliculi, since they lie at the heart of eye and head movements, which are very important for visual function and especially significant for Lucy's development. Like many structures in the lower brain, the colliculi are capable of controlling simple reflexes without cortical involvement. For example, the superior colliculus causes us to turn our eyes, heads and bodies towards a sudden visual movement and, in conjunction with the inferior colliculus, is involved in causing a similar 'orienting response' to a sudden sound. The cortex has connections to the superior colliculus from several of its maps and hence can supervene, using these pre-existing functions to help with the processing of more complex eye movements, such as those we make voluntarily.

Back up in the cortex, the naming of parts gets rather more confusing, since most of the labels were applied a long time ago, with great fervour and little regard for functional significance. Anything that has a vague texture to it, or bends in a manner reminiscent of some sea creature, will be given its own name, regardless of whether it is a thing in its own right or simply a consequence of the way the growing brain gets crushed within the skull. For our purposes, the only divisions worth mentioning are the major lobes of the cerebrum. To maximise the amount of cortical surface on the two cerebral hemispheres, the brain has evolved a convoluted structure. The individual convolutions are called gyri, and the gaps between them sulci (needless to say, every gyrus and sulcus has its own unique name), but every now and then there is a much deeper cleft, or fissure, that creates a major division in the cortical surface. The structures thus divided correspond to main branches in the cauliflower floret of the brain and are called lobes. Because these are fundamental subdivisions in the wiring diagram of the brain, each pair of lobes tends to have a characteristic broad function. At the back we have the two occipital lobes, which are primarily concerned with vision. The visual stream seems to split from here into two main branches, one of which enters the temporal lobes on the rear flanks of the head. These are implicated in the recognition and memory of objects, places, facial expressions and suchlike. The other stream penetrates the parietal lobes at the top of the head, which tend to handle spatial relationships, for example the conversion between eye-relative coordinates and arm-centred coordinates involved in grasping a visual object. At the front, conveniently, come the frontal lobes, and at the rearmost edges of these we find the primary motor maps, which are the final cortical stage in the control of deliberate muscle action. Spreading generally forward from this point, the frontal lobes seem to handle ever more abstract, delayed and strategic levels of volition, up to and including the elements involved in maintaining a sense of personal responsibility and pursuing long-term goals.

And that brings us rather breathlessly to the end of our whistle-stop tour of neuroanatomy. My aim was simply to introduce you to some of the main structures that lie between your ears, because these contain the functionality that somehow needs to be replicated inside Lucy's head too. The point of this project is not to slavishly copy these various structures, but to try to grasp the spirit of the design – the fundamental engineering principles that lie behind their internal structure and interconnections. Lucy's brain needn't look much like yours or mine, but it must at least draw inspiration from it. What else is there to go on?

Incidentally, the brains of human beings are fundamentally similar to those of all mammals, not just orang-utans and gorillas but even such relative dimwits as the hedgehog. The main difference seems to be that primates and humans have more cortical maps than the more intellectually primitive mammalian species. Saying that we have bigger brains is also true but misleading; it seems to be the number of stages (or perhaps types) of processing that lie between the primary sensory and primary motor areas, which marks us out as different, and having more stages of processing obviously requires more neural real estate. Humans, despite our deeply held prejudices, are not radically different from other mammals, and certainly not from other primates; in fact the differences are so small that most experimental neuroscience is performed on monkeys and yet is assumed to apply to us without much qualification.

For all I know, however, there may still be one exception to this rule, since, despite my best attempts to interview them on the topic, I have yet to discover what it is that goes on inside the brains of journalists.

CHAPTER TEN

Laughing pig curtains

When I woke up this morning three mischievous chimpanzees were staring at me from the window. Yesterday it was an avuncular gorilla. This recent shift towards primates is an unsettling trend that I suppose might have something to do with Lucy being loosely modelled on an orang-utan – I've got apes on the brain. Until recently, many of the faces I've seen at the window have been human, although the major culprit is a laughing pig.

When we moved house recently, the previous owners left behind some curtains, hanging either side of french windows that open on to a balcony. To me, these curtains are absolutely hideous but they are altogether too fascinating to part with. They are of a bold floral design, or at least, logic suggests that the artist probably had it in mind to draw flowers, and for all I know was very pleased with the result, but what I see in the garish patterns are faces – dozens of them.

It is well known that we have a region in each of the temporal lobes of our brain that is deeply implicated in the recognition and memory of faces. Faces are so important to social animals like us that it isn't a big surprise to find that we are exquisitely sensitive to face-like patterns, although I wouldn't want you to conclude from this that we have 'evolved a part of our brains specifically for recognising faces', because I think that might be a serious misapprehension. Anyway, it isn't the fact that they are faces that interests me here, but the fact that they are not flowers. I don't see what is actually there in the patterns on my curtains, but what I am already predisposed to think is there. Clearly I see with my mind, not with my eyes.

Mistaking a pink rose for porcine amusement is just an interesting and harmless optical illusion. I'm not as mad as I seem, or if I am then it has nothing whatsoever to do with my bedroom curtains. My visual system is simply playing tricks on me by forming false expectations. But the very existence of these expectations, and the way they overrule the evidence of my eyes runs much, much deeper than silly optical illusions: it is the very essence of my existence. I am the story I tell myself.

I hardly know where to begin with this subject, because expectations (false or otherwise) are so central to our being and yet masquerade under so many pseudonyms that the phenomenon almost gets lost in the crowd. I think I'll start by talking about signal delays, because they may be the root problem that the evolving brain had to face and from which so much else has arisen.

Imagine a bird flies past your window, and you catch sight of it and track it with your gaze. The signals from your retina take more than a twentieth of a second to get to the first parts of your brain that control eye movements and then the commands take about the same time again to get back to your eye muscles. So, if your brain directs your gaze to where the bird was when the image of it hit your eye, it will no longer be there by the time you look. Trying to track it reactively like this would be futile, and the bird would forever be running ahead of your gaze. But it doesn't. As long as you've had a short time to assess its trajectory, you can track it precisely and automatically. You must therefore be predicting its future position, not simply reacting to its present position.

You might object (with some justification) that the perception that you are accurately tracking the bird could simply be an illusion – after all, the bird may no longer be where your eyes are pointing, but the image reaching your brain is also out of date by a similar amount. Perhaps your eyes are looking at the wrong spot, but you still see the bird there because that was where it was when the image hit your eye. This troubled me too. So we'll start an imaginary clock with the bird at position 1 and think the problem through more carefully. Let's suppose it takes one whole tick of the clock for the signal to reach your brain and another for your eye muscles to receive a command in response. After the first tick the bird will have reached position 2 but your brain will have just received an image of the bird at position 1, so you will start to direct your eyes to where you think it is (position 1). By the time your eyes get to position 1 the bird is now at position 3. Meanwhile, you are receiving the image that hit your eye one tick ago, which started out when the bird was still at position 2. You start to look towards position 2, and by the time you get there the bird appears to be at position 3 and is actually at position 4. It just doesn't work out – only half the delay has been accounted for; your perception would always be that the bird remains one position beyond your gaze and the reality would be that it is two positions beyond this point.

If you want to prove to yourself that this really isn't due to an illusion, you have my permission to take out a gun and shoot the bird. For this to succeed, your brain must unquestionably compensate for the delay in the incoming signal and anticipate the time it will take for the message to get to your muscles and adjust your aim. You need to be aiming the gun towards where the bird will be when the bullet comes up to meet it, or you will miss. Telling me that the bird only thinks it has been shot is stretching the objection a bit far!

These signal delays are a huge problem for large, fast-moving (and very edible) creatures like us. If the object is only a bird, what do you care about a small tracking error like this? But suppose it is a lion. We can't simply react to incoming stimuli now. We must constantly predict the future and act in advance of it, or things will happen to us before we are able to respond. Being swallowed is not something one can usefully react to. If there is one word that sums up more than anything else what brains are for, that word is *prediction*. Some of these predictions are relatively straightforward, such as tracking a moving target (there is some evidence that moving shapes produce wave-like disturbances in the retina itself, which may mean that the first step in the compensation process happens long before the signal reaches the brain) but prediction is a far broader and deeper aspect of brain function than this.

Both cognitive science and AI have a tendency (and sometimes even a zealous fervour) to treat intelligence as a reactive mechanism. The reason for this probably goes back to the rise of behaviourism in the middle part of the twentieth century. Behaviourism stems from the perfectly reasonable observation that internal mental states such as thoughts and feelings are (or at least were, before functional brain imaging came along) inaccessible to us. Our theories about them must therefore be inferential and perhaps untestable, and hence it was concluded by many scientists that they are not proper objects for scientific study. As a warning to immoderate mentalists this may have been reasonable, but somehow caution turned rapidly into dogma, and even the mere mention of internal mental states became taboo.

The behaviourists became the Thought Police of the cognitive science world. Even today, some people get very hot under the collar about it, either because they fear the Spanish Inquisition will arrive on the doorstep intent on wrecking their reputation or because they genuinely believe that the reactive, conditional response theories of intelligence are the correct ones.

Obviously I don't want to get myself into any trouble here, so let me give my opinion delicately, diplomatically and cautiously. I think the 'reactive' point of view sucks. It is complete gibberish. It is utter bilge. For any fast-moving animal larger than an insect, reactive mechanisms are completely inadequate for most tasks performed by the central nervous system, which is therefore obliged to support internal mental states that correspond (either directly or indirectly) to an active model of the outside world, which is capable of producing predictive judgements that can drive proactive and anticipatory behaviour.

Such a viewpoint involves words and concepts such as 'internal representation', 'mental model' and sometimes even 'symbol manipulation'. Such phrases are anathemas to many people in modern AI, and merely uttering them seems to mark me out as a member of the older Symbolic AI movement – an approach that is now, quite rightly, largely discredited (even though nobody seems to have told many of the researchers). Let me reassure you that I am absolutely not an advocate of Symbolic AI, nor that my disagreement with the more modern reactive approach to AI implies that I belong to a diametrically opposed camp. But to state my case in depth would be inappropriate in a popular science book, and I doubt if anyone outside the academic priesthood would care that much anyway, so I shall take it that you have not been driven off in a huff by such blasphemies and are happy to read on. All you need to know is that at the heart of my work with Lucy is an assumption that the brain is fundamentally a prediction machine. What Lucy's brain will ultimately do (I hope) is create a mental narrative about the world; a constantly updated explanation of what is happening now and therefore what might happen next. If I get it right she will have more direct and meaningful access to this internal narrative than she does to the information entering her senses. What she will see is what she expects to see, not what her senses are belatedly telling her. She, like us, will live not in the real world, but in a virtual world in her head. She may even begin to see pigs in the curtains.

Let me illustrate the operation of this mental narrative with a very short story. When I was about thirteen, my family and I went for a holiday in a caravan on the south coast of England. On our first day the weather wasn't great, so we walked about a mile along the beach to the shops and bought a jigsaw puzzle to keep us amused. Sadly, when we returned and tried to assemble it we discovered one piece of the jigsaw was missing. It must have dropped out of the box somewhere on our way home. Oh well, never mind; there was nothing we could do about it now. The sun had come out, so we went to look for interesting pebbles instead. Suddenly, about half-way along the beach, something bright and attractive amid the millions of pebbles caught my sister's eye. She bent down to pick it up, and there in her hand, soggy from the recent rain, was the ...

Aha! You've just performed a miracle. I haven't told you what happened next and yet I'm sure you've already seen it take place in your mind's eye. As you read my tale you reconstructed, in your imagination, the events I was relating, but your reconstruction always kept slightly ahead of the game. I bet you simply couldn't stop yourself from seeing the jigsaw piece in my sister's hand. But until now I hadn't told you that this was what she found. The tale was new and unfamiliar to you, so you couldn't possibly have known from past experience how it would end. You therefore must have seen into the future. Exactly as I predicted you would. So are you clairvoyant? Am I? No, of course not. We don't really know what is going to happen in the future, only what is likely to happen. Nevertheless, it's an amazing skill and it lies at the heart of what it means to be intelligent. The fact that you simply couldn't stop yourself from doing it also says something about consciousness. I want Lucy to share this remarkable ability to predict the future. In fact this is the whole point of her existence.

Anticipating the next step in a written story is an example of a very high-level predictive process, as is forming a long-term plan or rehearsing a job interview in your mind so that you can consider your responses in advance. Guessing what someone else will do next by placing yourself in their shoes is perhaps the most sophisticated of all predictive models. Between these high-level predictions and the very low-level compensations for fixed signal delays that enable us to go bird watching comes a whole host of other things, all of which involve keeping track of what is happening and using this information to predict the future, or sometimes to fill in for missing or ambiguous information and thus essentially to predict the present.

Many optical illusions illustrate how we use context to resolve ambiguity or compensate for limited information in this way (including the illusion of seeing laughing pigs and chimpanzees in bedroom curtains). Below is a rather nice optical illusion, which demonstrates that what we see can depend on quite subtle contextual clues. The two horizontal lines are actually identical, yet if we are shown them at different times we are likely to interpret one picture as a line drawn through a column of Xs and the other as a pair of arrows, simply because of the surrounding context. It seems to me that a prediction at one level of abstraction often provides the context for making predictions at a lower level, and the arrival of anticipated (or in a slightly different way, unanticipated) sensory events at the low level then confirms or moulds your belief about events at the higher level. In the optical illusion, the low-level visual information surrounding each horizontal line gave birth to two different high-level interpretations of the kinds of symbol that might be present, and each provided a different context for disambiguating the meaning of the (identical) horizontal lines.

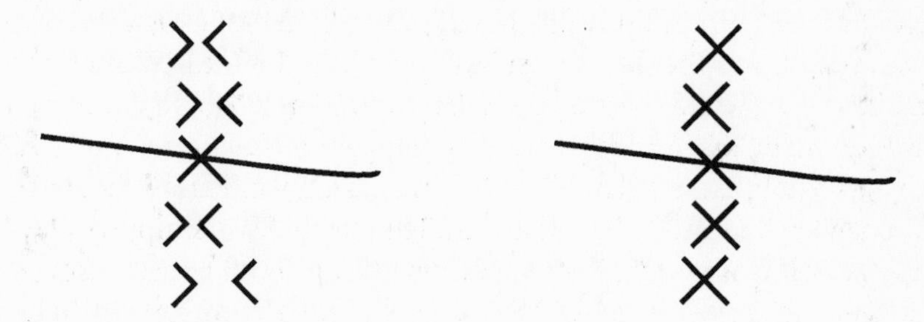

Figure 14 Arrows and crosses illusion

Even the process of performing a physical action seems to me to be a part of our inner narrative, only now we are seeing the future we intend, rather than the future we passively expect. The psychologist William James called this process 'ideomotor action'. The most compelling example I know of for such goal-oriented control of movement is flying a light aircraft. I realise this is not an analogy that will be familiar to everyone but, as you've already demonstrated, you have a powerful imagination, so go with it. To fly a small plane from one place to another, the best way is not to wrestle with the controls or think about what you are doing but simply to imagine the result you want to achieve, and then it happens, almost like magic. The movements you make to the joystick and rudder pedals are often imperceptibly small, completely intuitive and utterly beyond your conscious control, yet you tell a story inside your head about where you intend the aircraft to be and your muscles automatically comply with your wishes to make the story happen. It's not quite the same thing, but if the flying example doesn't work for you, imagine driving a car through a very narrow gap – the trick is simply to wish yourself through to the other side. On no account should you think about the steps you actually have to take to achieve it, or you'll come a cropper. People who advocate 'positive thinking' are taking this idea to extremes, to the extent that they sometimes appear to think that merely stating a goal and believing it will happen is enough to ensure that it does. Sometimes they're right. Ideomotor control extends beyond the body, beyond the extended body of a plane or a car and influences our course through life in general.

Most of these predictions and their consequences are managed unconsciously. Consciousness of the external world doesn't usually come into play in our active lives unless something goes wrong, or we are experiencing something new or relatively unfamiliar. In the mornings, I can get out of bed, get dressed, washed and shaved, feed my tropical fish and make coffee without being the slightest bit conscious throughout the whole episode. Sometimes I don't even think I choose to do these things – getting out of bed is invariably something that happens to me, rather than something I consciously initiate. The only time consciousness plays a part is when I discover I've put fish food in my coffee cup or I'm trying to clean my teeth with a teaspoon.

When the process is so unfamiliar or difficult or imperative that it requires the full attention of our conscious minds, we say that we are thinking, reasoning, planning or imagining. When it is automatic as a result of previous learning we call it instinct or reflex, if we call it anything at all. Whether these conscious and unconscious predictive mechanisms are at opposite ends of a continuum or fundamentally different things is something I'm going to have to think about. For the moment I'm assuming that they are at least strongly connected, and I'll refer to this internal predictive model or models in general terms as 'imagination'. I want Lucy to be imbued with an imagination.

One thing that strikes me as notable about the narrative that's going on inside our conscious minds is that it can be maintained and adjusted quite easily, but takes a considerable amount of time and cognitive effort to build up in the first place. When we come out of a general anaesthetic, or even when we wake up in a strange environment, it can take us quite some time to make sense of what is going on. Thereafter, as long as our experiences remain in approximate lockstep with our expectations, we need only to tweak the model here and there in order to keep it in tune with events. It's as if the model has a kind of inertia: it takes time to get it up to speed, but then it runs ahead of us with a certain degree of conviction, and this is how it manages to keep one step ahead of reality. We can nudge it gently around bends to deal with slightly unanticipated events, but if the world changes too radically, we're apt to believe for a while that we can see laughing pigs at the window, until our model gets back on track.

It seems to me that many aspects of the brain's activity can be described in terms of keeping track of an internal story or model of what's happening in the outside world, which we use to estimate what's likely to happen to us next, to prepare ourselves in good time for taking action. But I don't mean that this is necessarily a story in words – thinking in language is one of the slowest kinds of thought. Nor is it likely to be a centralised or even coherent story. I think we have to imagine this model existing as an approximate and variable consensus among many smaller stories being told in various parts of the brain, each in the 'language' that this part speaks. So our visual system is keeping track of what we are seeing and therefore what we expect to see, and it does so in terms of visual expectations, such as lines and edges (or the future position of a bird flying past the window). Meanwhile, our motor system is telling itself stories about where our limbs are going to be in a few milliseconds' time, narrated in the language of joints and muscles. It anticipates their future disposition well enough to be able to pull on the brakes in time to stop them precisely where we intended them to stop. The spoken stories that come out of our mouths might exist beforehand in the form of words, but not necessarily as discrete symbols, just sequences of mouth movements and patterns of sounds. Even then, most of the time we have to wait to hear ourselves speak before we know what we were going to say. The parts of the narrative that keep just ahead of what we are hearing or reading (as in the jigsaw anecdote above) are not verbal either – you will have 'seen' the jigsaw piece and maybe, had I been more descriptive, have heard the sea whispering across the pebbles and smelled the salt air.

When all is well, reality and the storyline remain in approximate lockstep. The various sub-plots being told in different parts of our brains link up and more or less agree with each other, and the things we think are about to happen, or are likely to happen, do happen. Except, of course, when we dream or daydream, speculate or make plans, when our inner story is allowed to drift, released from its ties to reality. This fact is quite illuminating in itself, because if our imaginary stories are to unfold realistically in the absence of sensory input, certain things need to be true about our brains. It's no good if you attempt to imagine yourself walking but your brain refuses to admit that your imaginary legs are moving, simply because your real ones aren't. Moreover, imaginary movements have to take place in real time, so our models must not only anticipate what happens next but when it will happen. But I'm getting ahead of myself again.

The main thing I've been thinking is that perhaps, just perhaps, many or most of these predictive capacities are really examples of the same mechanism. Perhaps beliefs, desires, intentions, plans, perceptual hypotheses, anticipations and delay compensations are really all examples of the same process, but when they are handled by very abstract and long-term regions of the brain we call them plans, when they are occurring in early sensory regions of the brain we refer to them as sensory hypotheses and when they occur in motor regions we call them intentions. If so, then my problem is much simpler than it might otherwise have been. Get one mechanism right and I have them all, at least in principle. What's more, I can draw on many apparently disparate examples of internal mental activity for clues about what their common framework might be. This can't be the whole truth, and I'm sure that there are many special cases, in which predictive mechanisms have evolved to handle specific tasks in unique ways. But seeing the brain as a general-purpose prediction machine, and seeing many predictive capacities, from planning through to muscle sequencing, as examples of the same basic mechanism, does at least fit with the apparent uniformity of the cerebral cortex. Such a thought doesn't necessarily get me very far with building Lucy, but at least it's a handle to pick the problem up with. It helps get the ball rolling. I wonder what I'll think of next . . .

CHAPTER ELEVEN

Learning to fly

Before Lucy became an orang-utan she was an eagle (see plate 5). That is to say, before I started building Lucy as she is now, I had been working with a two and a half metre wing-span model glider, which I was hoping would be able to learn for itself how to fly. Unfortunately, I hadn't reckoned with the dreadful English weather, which gave me precious few opportunities to test it. The model also got rather too heavy, thanks to all the computer equipment on board, and since heavy aircraft fly very quickly, I started to have nightmares about what would happen if it hit someone. In the end, the glider threw itself ignominiously into the ground in front of a television crew, and I decided that enough was enough. In any case my thoughts had moved on and I needed a rather different test bed to try out my newer ideas.

Learning to fly is a perfect challenge in many ways, and very unlike most robotic environments. For one thing, it is often assumed that artificial intelligence systems (and for that matter natural ones) will make discrete, all-or-nothing decisions: turn left; move forward; Queen to Knight's Pawn 3. To fly a light aircraft, there are only a small number of things you need to twiddle with most of the time: a joystick, a rudder and perhaps a throttle. Flying is not about deciding *what* to do, but rather deciding *how much* to do it – how much to push the stick forward, how much pressure to put on the left rudder pedal, how high to climb. For real animals, including humans, most activity is like this too. We may decide explicitly to go shopping, but to execute that decision we need to steer a car, navigate the supermarket aisles and so on. Even when we seem to be making a discrete decision it could be simply that some continuously variable tendency has passed through a critical threshold and become a 'decision', in much the same way that water suddenly boils as the temperature gradually rises.

In many respects, trying to keep a glider aloft by keeping it straight and level, deciding whether and how long to stay in a thermal or seeking out the centre of lift, are more characteristic of everyday animal life than the sorts of artificial decision trees and open-loop plans beloved of many AI researchers. The other, far more important reason for working with a model glider was that the concept of a servo loop turned out to be the stimulus for my first piece of real inspiration in my quest to understand how 'imagination' might be implemented in a neural network, and it just so happens that model gliders are controlled by servos. Let me try to explain.

As you'll remember from chapter 5, a servomotor is a device that seeks out a desired rotor angle. You tell the servo which way you want it to point by sending it an appropriate signal, and it will make whatever movements it finds are necessary in order to achieve that goal and maintain it, even if it meets unexpected resistance on the way. In a model glider, servomotors control the position of the ailerons (the flaps on the wings that cause the aircraft to roll), the elevator (the horizontal flaps on the tail that cause it to pitch up and down) and the rudder (the vertical flap on the tail that causes the glider to yaw from side to side). The pilot on the ground twiddles a pair of joysticks, which send signals via a radio transmitter to the servos in the aircraft, where they drive the flaps to the desired angles. The angle of the joystick dictates the angle of a given flap, regardless of how much wind resistance and friction the servo must overcome in order to get it there.

The specification of these desired flap angles by means of a joystick can be thought of as 'first-order desires', for reasons I'm about to explain, and their consequence is to cause the aircraft to *accelerate* in a particular rotational direction. Move the aileron joystick to the left and the aircraft will start to roll to the left at an increasing rate. Leave the stick in this position and the roll will get faster and faster. To make an aircraft roll at a *constant* rate, we must therefore push the stick over to one side, wait until it is rolling at the rate we desire, and then centre the stick again. In practice things aren't quite this tidy, but my description will do for the sake of argument. If the air is the slightest bit turbulent, we would have to keep moving the stick slightly in both directions to compensate for the turbulence and maintain a constant rate of roll.

If the desired flap angle is regarded as a first-order desire, then the desired rate of roll can be thought of as a second-order desire. To reach

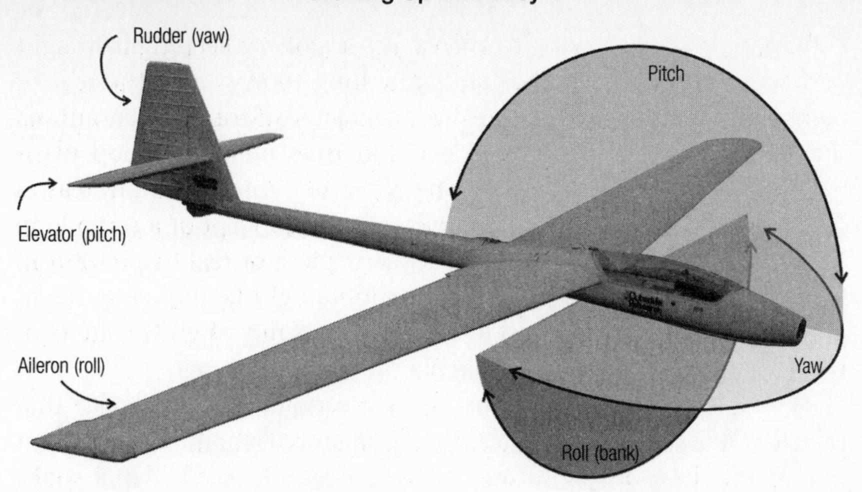

Figure 15 The three axes of rotation and the control surfaces that affect them

and maintain a desired rate of roll, we have to keep *adjusting* the rate of acceleration, and hence the flap angle. In making this goal-seeking adjustment we are behaving exactly like a servo ourselves. We are constantly adjusting our instructions to the physical servo to achieve our own goal, while the physical servo constantly adjusts the force it applies to the aileron flap to try and achieve each new goal that we set it.

Consequently, if we were to attach an electronic or software equivalent of a servomotor to the input of the physical servo that controls the glider's ailerons, then we wouldn't have to do all this fiddling about ourselves. We could simply dial in a desired *rate* of roll to this 'virtual' servo, and it would automatically send the necessary sequence of signals to the actual servomotor in order to adjust the aileron angle in a way that achieves our aim. The joystick would no longer control the raw flap position but would directly control the rate of roll.

If we take this idea one step further, to the third-order level, you will, I hope, begin to see how neat it is. If the first order is the desired angle of flap and the second order is the desired rate of roll, then the third order must be the angle of bank of the aircraft. By having three servos chained together, two of them virtual and one physical, we can fly the aircraft simply by specifying what angle of bank we want to achieve. Push the joystick to the forty-five degree position and the third-order servo will tell the second-order servo that it initially desires a rapid rate of roll to the left. The second-order servo will attempt to achieve this by telling the first-order one to accelerate the aircraft's rate of roll from zero to leftwards. The physical servo will then alter the flap position to attain and maintain this rate of acceleration, and once the aircraft is rolling at the desired speed, the second-order servo will tell the first-order one to ease off the acceleration. The aircraft will now roll left at a constantly maintained rate, until the angle of bank starts to approach the desired forty-five degrees, whereupon the third-order servo will alter its signals to the second-order one so as to slow the rate of roll and eventually maintain this angle of bank.

In a way, these extra virtual servos are making the glider more 'intelligent'; enabling it to respond to higher-level commands from the joystick, such as 'tilt over to forty-five degrees and stay there'. Before I try to relate this notion to imagination, prediction and real intelligence, there are several features of the analogy worth pointing out.

The first thing to note is that each servo in the chain requires different sensory information. A servo attains its desired state by comparing this with its actual state and deciding how it should move in order to minimise the difference between the two. The sensory feedback to the first-order, physical servo comes directly from a sensor underneath the motor shaft. The servo is given the desired shaft angle and the sensor tells it the actual shaft angle. From these it can compute how much and in which direction to apply current to the motor in order to bring the actual angle in line with the desired one. But the second-order servo needs to know how rapidly the aircraft is rolling, in order to compare this to the desired rate of roll and work out what flap angle to request from the physical servo. Likewise, the third-order servo needs to know the current angle of bank of the aircraft, in order to decide what rate and direction of roll will bring this in line with the desired state.

So a hierarchy of servos implies a hierarchy of sensory systems. Some of these are relatively easy to implement, but many would require some sensory processing. Determining the current angle of bank can be performed by an accelerometer or a pendulum, which detects the direction of gravity, but if the aircraft is accelerating at all, in any direction, the reading will be inaccurate (because gravity and acceleration are equivalent, as Newton pointed out), so some extra computation is required. Alternatively, the angle of bank could be measured optically, using the horizon as a reference, in which case some kind of simple visual processing would be required.

The second thing to notice is that this system is fully reversible. If we desire an angle of bank of zero (in other words, if we ask the aircraft to fly straight and level), the servos will chatter to each other in such a way as to achieve this state. If the aircraft then encounters turbulence, the servos will automatically react to the difference between desired and actual states in order to bring it back to straight and level again, without us having to do anything at all. It doesn't make any difference whether it is the desire or the environment that changes, because the servo simply attempts to minimise the difference between the two. This, of course, is how an autopilot works. It is also very reminiscent of the way our minds work. A decision might be a consequence of a change in our plans, or it might be driven by a change in our external circumstances; in either case we require a similar response to bring our external and internal worlds back into line.

The third feature stems directly from the previous one. This is the fact that otherwise completely separate chains of servos can become loosely coupled to each other through the environment. The reason pilots bank their aircraft over to one side is to turn them. As it banks over, a plane will start to yaw towards its lower wing, and so begins to turn. However, this rolling and yawing has at least one undesirable consequence. Because the banked-over aircraft no longer has its wings level, they don't provide as much vertical lift any more, and so the nose drops. To fly a neat turn you need to hold the nose up using the elevator, or you will quickly find yourself entering a spiral dive. The obvious engineering approach would be to link the elevator servo directly to the chain of servos controlling the ailerons, but in fact no such link is needed. If we imagine the aircraft has a second, completely independent chain of servos that do for pitch angle (using the elevator) exactly what the first chain does for bank angle (via the ailerons), then these two chains will automatically interact via the environment.

Imagine the plane has been told to fly straight and level, and so has zero angles of pitch and bank. If we then bank the aircraft over to begin a turn, the nose will start to drop. But the elevator servo chain has been explicitly told to keep the pitch angle at zero, and so it will automatically respond to the dropping nose by applying a little up-elevator. The servo chains don't need to talk directly to each other, and we don't need to lift a finger. Such behaviour is extremely common in living things too. If we decide to lift one leg off the floor, we don't simultaneously have to think to alter our balance to bring our centre of gravity over our other leg to avoid falling over. It happens automatically, thanks to a set of loosely coupled servo systems.

The fourth insight comes when you think about extending these chains of servos still further. The third order in the roll axis chain allowed us to set a desired angle of bank. The angle of bank defines the aircraft's rate of turn, and so a logical fourth order in the chain would be the facility to set a desired heading. By simply giving our glider four chained servos with appropriate sensory inputs, we can flex a joystick to tell it the compass direction in which we want it to fly. It will automatically roll into a turn and roll out again when it reaches the appropriate heading. If it then gets knocked off course, it will turn left and right to compensate. Beyond this level there are no extra orders of control in the roll axis that have much direct usefulness, but one of the main reasons you turn a glider is to bring it into a source of lift and keep it there, so turning does have a consequence for the maintenance of altitude. It may be that we wish to specify a variety of target altitudes to reach but in practice a glider generally wants to be as high as possible, which means that the desired altitude will be permanently set at infinity.

When you find such fixed desires in animals they are called 'drives'. We perpetually desire to be neither hungry nor over-full, to be warm but not hot, to have companionship without being overcrowded and so on. At the head of its chains of servos, the glider might have a fixed desire to be as far away from the ground as possible, and another to be as close to home as possible. All the other servos will then fidget and chatter to each other, with the collective aim of ensuring that these fixed goals are achieved and maintained.

To do this requires learning. It may be mathematically obvious how much you need to move a motor to bring its rotor to a given desired angle, but it is less obvious exactly how you should alter pitch angle to control an aircraft's speed, and extremely difficult to work out from first principles which way and how much to turn an aircraft in order to seek out and stay within a nebulous patch of rising air. To be truly intelligent, the servos must therefore be capable of working this out for themselves, through trial and error. Ideally, the servos would even work out *which* other servos they should be influencing, in order to minimise the difference between their own target and actual states. In other words, the servos would be capable of both wiring themselves up into chains or networks, and altering their own internal rules. Interestingly, the servos need never know what part they play in the whole enterprise. The only thing that need concern them is what they can do with their output signal that results in keeping their target input and sensory input as similar as possible.

To see why this way of looking at things really excited me, forget about simple creatures like gliders and think about how it might be implemented in something as complex as you or me. Even the simplest actions of animals (at least land animals) require extremely complex output signals. To turn an aircraft requires a few gentle movements on the stick and rudder but for me to turn requires the complex interactions of dozens of muscles. Nevertheless, the basis of these muscle movements is servo action, trying to maintain specified forces or accelerations as conditions change. Just imagine what a maze of non-linear servos you would need to control a human being. It isn't necessarily impossible, just a great deal more complex than a glider. Imagine these servos spread out in a sheet, interconnected by nerve fibres. Imagine that each servo constitutes a column through the thickness of that sheet, and is made up from a cluster of interconnected neurons of various different types. Each column is a standard servo, but the internal structure of every servo is slightly different, because of the input-output relationship it has learned. This relationship enables it to minimise the difference between its target state and its present state, by sending new targets to other servos.

You can probably visualise how this sheet of servos would be forever adjusting itself, forever trying to minimise the difference between the thousands of actual states and their associated target states. No servo needs to know what part it plays in the scheme of things; it simply has to know how much to twitch in order to bring its actual (sensory) state into line with the target given to it by another servo and it does this by specifying new targets to one or more other servos, lower in the chain.

Now think about these target states and present states separately. Every servo has a present state – some numerical measure derived from senses that tells it how the world is. If we lay the servos out in a sheet then these present state values could be represented as vertical bars on a three-dimensional bar graph. The complete forest of vertical bars defines a surface: a landscape representing the total sensory state of

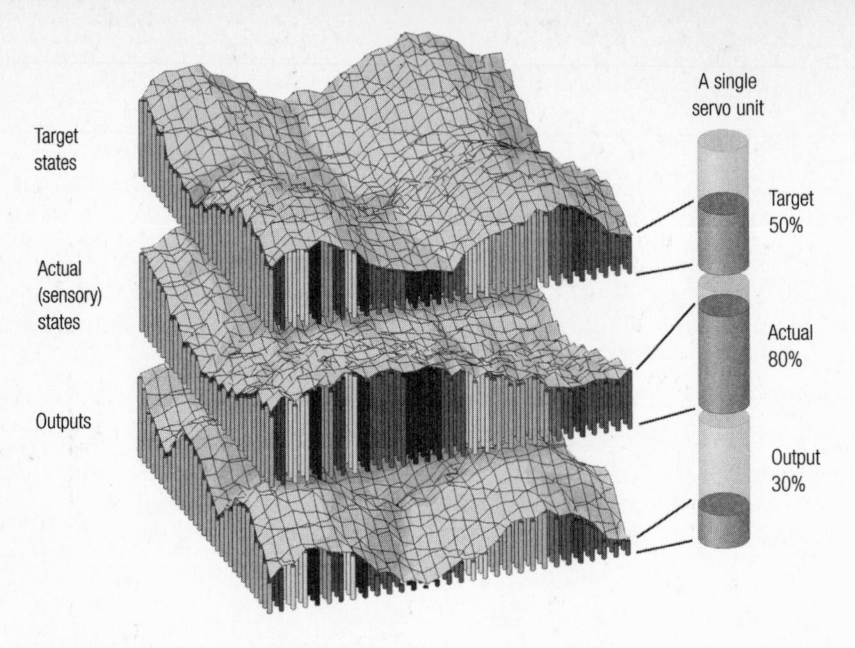

Figure 16 Array of 'servos' showing the Target, Actual and Output activity surfaces

the creature and, as the external world changes, this surface will buckle up and down. Similarly, the target states can be mapped out as another quite separate surface. The job of the creature's entire brain is to bring the sensory surface into line with the target surface.

In a completely 'happy' creature, these two surfaces would mesh perfectly but if the creature grew cold or hungry or tired or frightened, one or more of the bars would deviate, leaving a gap between its target and present state – a reflection of the creature's desire. To compensate for this and bring the creature back into its comfort zone, these discrepancies would need to cause changes to the target states of many other servos, bringing them temporarily away from equilibrium and causing still further changes in desire. A flurry of activity would ripple across the network, resulting eventually in changes to the target states of servos that are connected directly to the outside world (rather like the physical servomotors connected to the glider's flaps). The outputs of these servos would cause changes, not in the internal world of the creature's desires but in the external world: muscle movements that eventually alter the creature's sensory state in return.

At first these sensory changes will be primitive, but they will cul-

Lucy

The 'competition', demonstrating just how far AI is from replicating anything approaching primate level intelligence

Peta is the adopted 'granddaughter' of Gary Shapiro, VP of the Orangutan Foundation International. For more information see www.orangutan.org

One of Lucy's computer boards

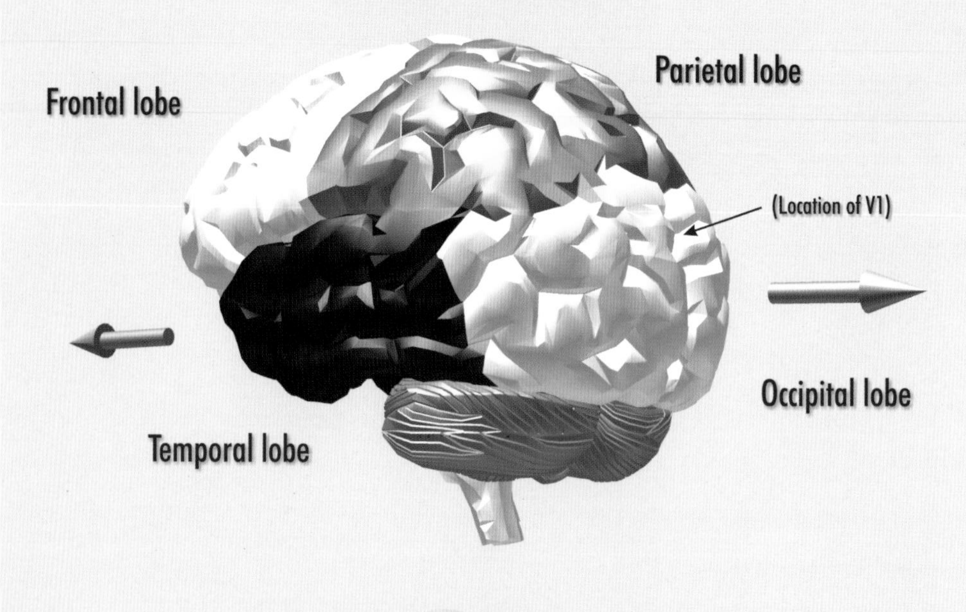

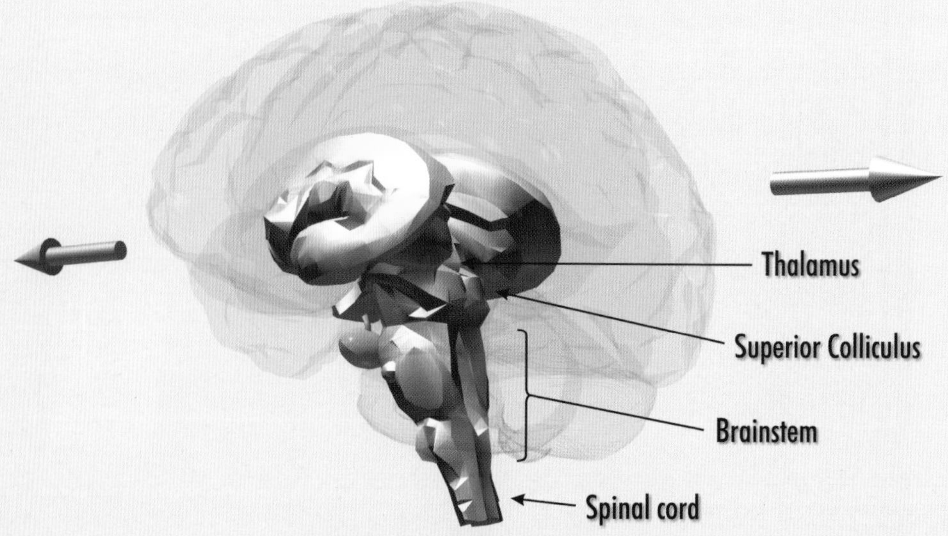

The cerebral cortex (top) and some of the major sub-cortical structures

Before Lucy was an orangutan, she was a model motor-glider

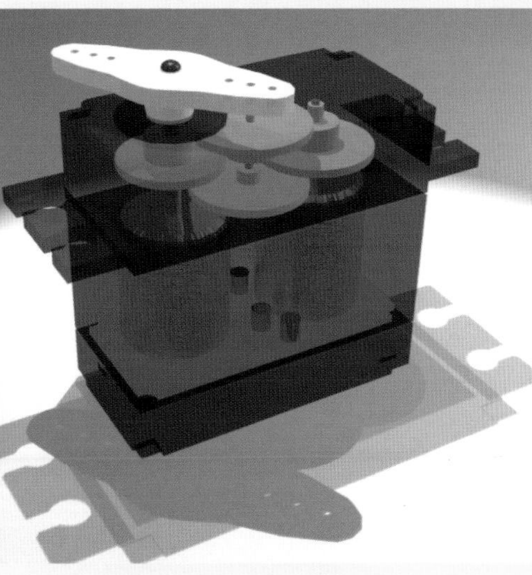

The growth of orientation detectors in Lucy's V1

Lucy staring at a display of moving stripes. She could learn equally well by looking around at the real world, but the computer display allowed me to control the conditions and try various experiments

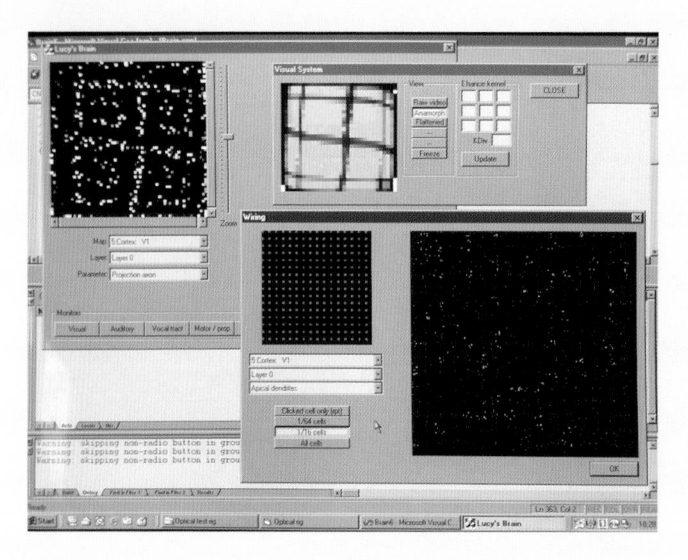

Some windows into Lucy's brain, showing the distorted view through her eyes (top centre), the firing of some of her V1 neurons (top left) and the patterns formed by dendrites in a small part of V1 (bottom left)

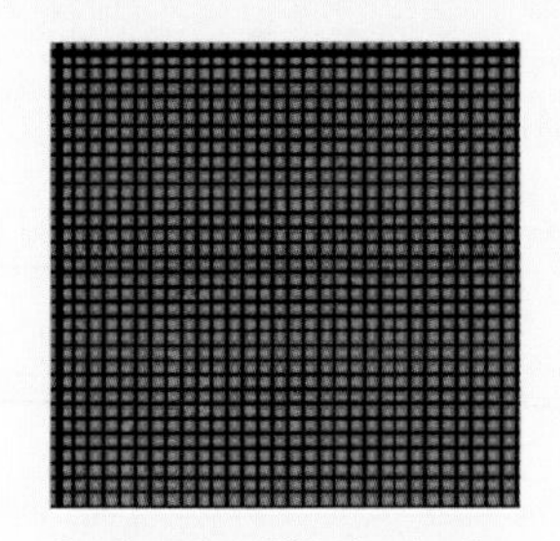

Small section of V1, showing the distribution of dendrites before Lucy's eyes have opened. Notice that all the cells have roughly circular receptive fields

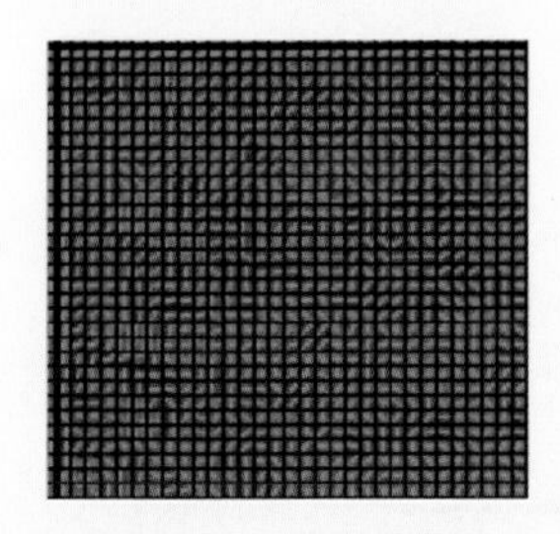

After exposure to moving lines the cells have become specialised to detect different orientations. Each small region of visual space tends to have a cluster of cells representing every possible orientation

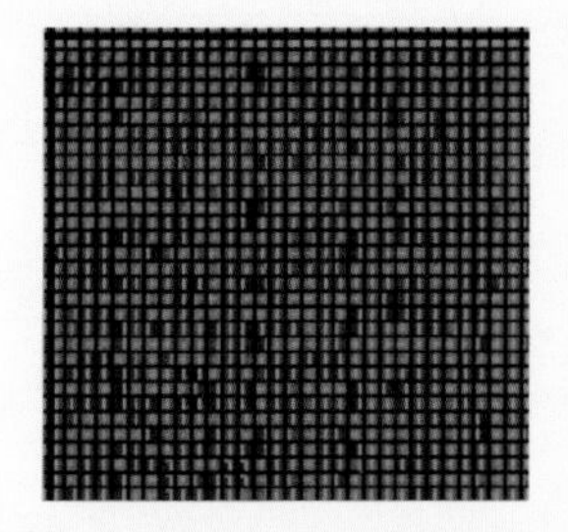

If Lucy is only shown vertical lines then receptive fields develop only one kind of directional specificity

After learning, this is the pattern of cell activity produced on the surface of V1 by a chequerboard pattern. The white dots represent cells that are stimulated by an edge to whose orientation they are particularly sensitive

The original image

has its edges enhanced

well-connected edges get selectively enhanced

the battle continues

until one surface momentarily wins

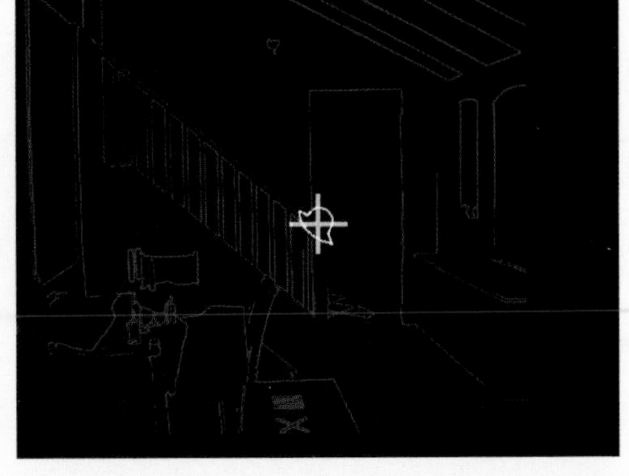

causing a saccade that centres the gaze on it

Simulation of how Lucy's V1 isolates a bounded surface and focuses her attention on it

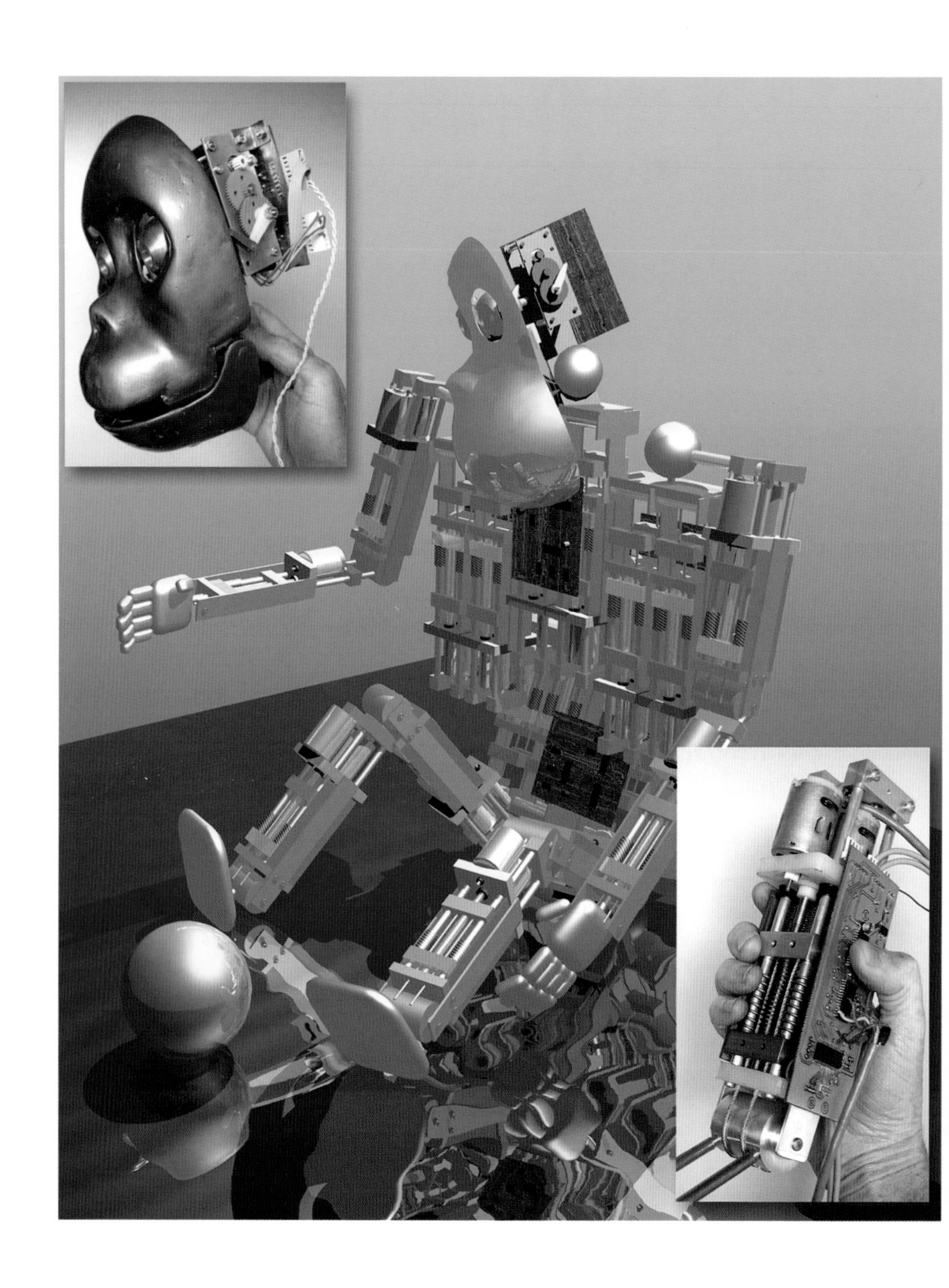

Lucy Mk II begins to take shape

minate in behaviour that reduces one or more drives, by locating food and eating it, for example. Eventually the creature will return to its comfort zone, with the two surfaces meshing once again, perhaps now in a different configuration. More likely, the surfaces never quite mesh, and the creature will perpetually seek an unattainable equilibrium, in which the difference between desire and actual state can be minimised but is never zero. This sounds more like reality: whenever we achieve something that we think will make us deliriously happy, we change in such a way that we become unhappy again and seek something else. Such is the dance of life.

To begin with, I thought that this notion of a sheet of servos might map very directly on to the structure of biological cortex, but now I realise that this is not the case. Nevertheless, it is still a very powerful metaphor, and seems to me to be a significantly different way of looking at what the brain is doing, compared to many existing computational models. Most importantly, I was struck by the idea of these two 'state surfaces', one sensory and the other connected with desires and intentions. If the sensory surface represents a physical state – a measure of how the creature *is*, then the desire surface is a *mental state* – a description of how the creature wants to be. Indeed it is a description of how it *expects* things to be, once the servos have had the chance to do their work and bring the sensory surface back into line. It is, therefore, a model of the anticipated future state of the world.

Is this just a flight of fancy, or are we starting to glimpse the outlines of an explanation for why we have an imagination?

CHAPTER TWELVE
Yin and yang

Intuitively, we tend to think of the brain as something akin to a production line. I call this the 'sausage machine model', because it implies that information flows in at one end of a tube and gets churned around inside the machine before being spewed out of the other end as some kind of action. We seem to use our senses to recognise the situation we are in, compare this with memory to decide how we should respond, and as a result, signals cascade into our motor cortex and initiate action. This is exactly the model I used for my previous attempt at artificial life, in the *Creatures* game, and it is implicit in most AI systems, whether based on neural networks or not. However, my dear Watson, the evidence for this case doesn't quite add up.

For one thing, how do we explain the fact that more than half of the signals in the cortex flow the wrong way – towards the senses, rather than from the senses towards regions that are 'deeper' in the brain? Many textbooks on the brain report this fact almost as an afterthought and ascribe to these connections 'some kind of feedback role'. But which direction is the feedback direction anyway? Surely what our senses do is provide feedback to us from our environment, monitoring our effects on the world? The temperature sensor in a central heating thermostat provides feedback about the thermostat's performance as it alters the temperature of the room by switching the heater on and off. So, if the bottom-up direction, from the senses inwards, should properly be described as the feedback direction, then these other fibres are in the feed-forward direction. But what, exactly, do they feed forward?

Several other things strike me as odd when I try to work out how the various cortical maps are inter-wired. It may be that I'm wrong about this. After all, I'm not a neuroscientist, and cortical wiring is extremely messy, complex and hard to pin down, but after drawing together a consensus of information from many neuroscience papers, I find myself with a general wiring scheme that completely fails to conform to the sausage machine model.

Cortical tissue, you will remember, is usually held to consist of six layers. Inputs from the senses characteristically arrive via the thalamus and enter primary sensory cortex in layer IV. Outputs arising from the pyramidal cells in layers II and III then tend to project forwards to secondary cortical maps, often entering them in layer IV in just the same way that raw sensory data does to the primary maps. Meanwhile, a set of returning fibres arises from layer VI of secondary cortex to end up predominantly affecting the cells in layer II of the primary map. Layer V, meanwhile, tends to send signals to sub-cortical areas concerned with triggering motor actions of one kind or another, and layer I can essentially be disregarded, as it contains very few neurons (it is from the maze of nerve fibres in layer I that many of the other layers receive signals).

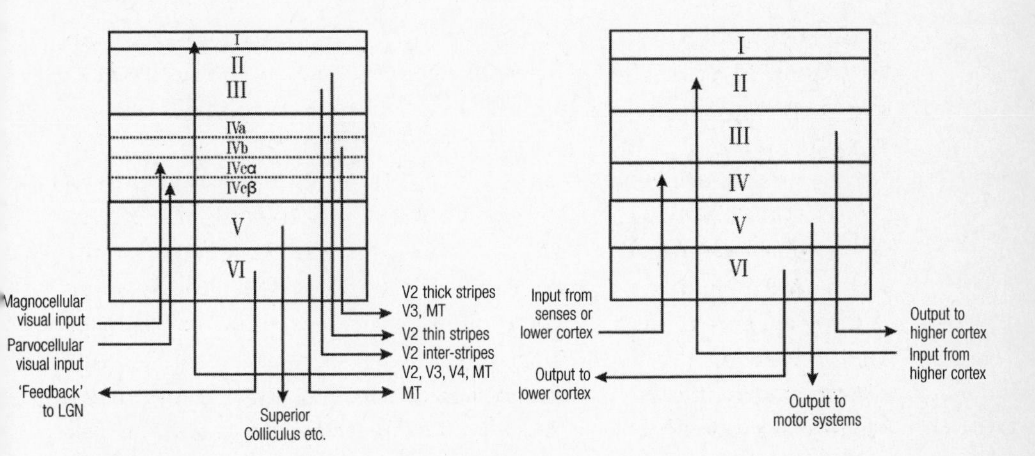

Figure 17 Major inputs and outputs of cortical area V1, plus a generalised map for cortex as a whole

This is all a bit confusing; so let me risk over-simplifying the structure still further. Let's ignore layer I, which consists largely of axons and dendrites. Let's also for the moment combine the large pyramidal cells in layers V and VI into one layer. Now we have a kind of club sandwich, with two layers made mostly from small granular cells (layers II and IV), interleaved by two layers made from pyramidal cells (III and V +VI).

Let's rename these idealised layers A, B, C, D, as in figure 18. Now we can see more clearly that there are two main chains of signals, one flowing in each direction. Sensory data enters in layer C and exits from layer B. It then passes on to the next cortical map, entering in layer C again. C Þ B Þ C Þ B ... The other chain flows out of one cortical map from layer D and enters the 'previous' (closer to the raw sensory input) map in layer A. D Þ A Þ D Þ A ...

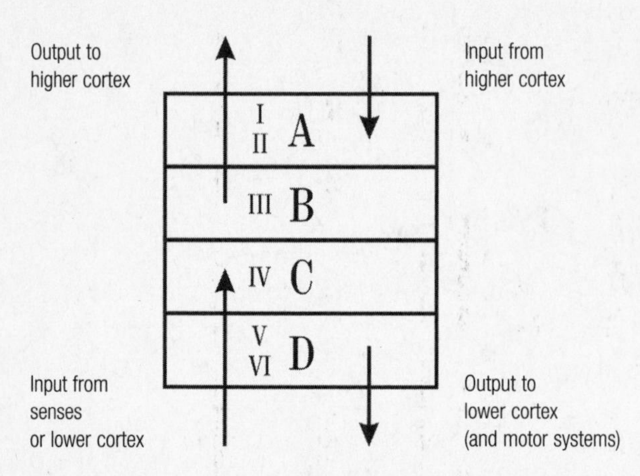

Figure 18 'Club sandwich' model of cortical connectivity

Now, if this simplification holds true, something is wrong. For the sausage machine model to make sense, we would expect to find that outputs to the muscles from motor cortex would arise from layer B. Sensory data enter the system in layer C and ripple through the network in the sequence CBCBCB. So the output of the sausage machine should come from layer B. But it doesn't. It arises from layer D. The signals that drive our voluntary muscles, for example, emerge from a cortical map known as the primary motor cortex, or M1, and they do so from layer v (layer D in my simplified scheme). Layer D is on what we have been regarding as the 'return path' through the cortex. Signals leave higher cortical areas from layer D and enter lower areas in layer A. ADADAD. We would expect this chain to flow backwards along the maps, emerging from the primary sensory end at layer D and projecting to something sub-cortical, and indeed this is what happens (from V1 it projects to the superior colliculus, of which more later) but this doesn't fit the sausage machine model at all.

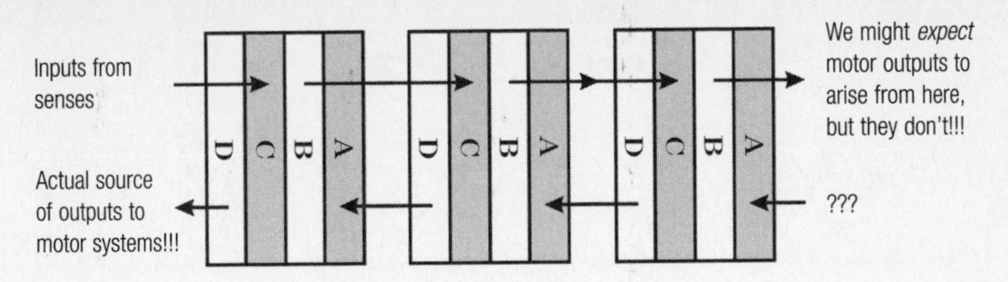

Figure 19 Several club sandwiches connected together

What are we to make of all this? Well, before I tell you what I make of it, I need to emphasise yet again that I may be wrong. I've checked this with a few people that I know but I may easily be misinterpreting or over-interpreting the neurological evidence. If you are a neuro-scientist you may be blowing a fuse with indignation at this point. I apologise for my ignorance and recommend that nobody rely on my description. If I were a professional scientist I would now feel obliged to spend some considerable time testing these interpretations and making sure my facts are straight. But I'm not a scientist. What I'm doing here is closer to the creative arts than it is to science. I'm playing hunches, exploring ideas and toying with possibilities. When one is looking for inspiration, it is often permissible to take something on trust that might be untrue; to see where it might lead, and use it as a weapon to beat one's way through the barriers imposed on us by our prevailing paradigm. I am setting these ideas up as a skyhook: something to hang other ideas on that might ultimately prove fruitful. These other ideas may well be right, and supported by other evidence, even if the means by which I arrive at them is wrong. My main goal is to build an intelligent android inspired by my understanding of the brain, not to produce an accurate model of how the human brain works. You have been warned.

Anyway, what I seem to perceive is two streams of signals running in opposite directions: CBCB and ADAD. Sensory data arrive in layer C, at the start of one stream, while motor signals arise from layer D, at the end of the other. The two seem to be on separate pathways. This does not look like a single pipeline, which takes raw sensory data in at one end, abstracts it, recognises it and spews motor signals out of the other. If anything, it looks like an outgoing signal and a returning echo, although since the creature's environment also bounces back the consequences of any motor actions to the brain via the senses, it is not obvious which of the two paths should be regarded as the echo – maybe a loop is a more accurate description.

To cut a long chain of reasoning short, it seems to me that the cortex is structured more like a tripod than a pipe. Visual signals enter the CBCB chain in map V1 of the occipital lobe but the outputs to our main body muscles do not arise from the end of the ADAD pathway of this same chain. Primary motor cortex is in the frontal lobe, so voluntary movement sounds like the end of a completely different pair of data streams. Much the same applies to the auditory system, the somatosensory system and so on. It is as if there are several distinct, bi-directional processing chains, arranged in a pyramid like sticks in a campfire. In truth the chains meet at many levels, so the real picture is much more messy, but the important thing is that both sensory and motor signals connect to the *same* end of the various chains of cortical maps, not to opposite ends.

So, primary visual cortex, V1, receives the sensory signals from the optic nerve at the start of a CBCB chain. But it must also be the end of an ADAD chain, and if so, I would also expect it to send signals to one or more of the motor systems of the body. We know that it doesn't send them to the main voluntary muscles, because these arise in M1, way off in the frontal lobes. So where do V1's motor signals go? Well, in fact they go to the superior colliculus – the structure concerned with controlling eye movements. Why do they do this? I shall return to this question with even more idle speculation in chapter 14, but for now it is enough to know that V1 obeys the rule and has motor outputs as well as sensory inputs. This alone is a striking fact, because V1 is usually described as a primary sensory map, but here we have evidence that it is also a primary motor map! The signals are there for everyone to see, but nobody seems to mention them, except in passing, perhaps because they don't make sense in terms of the sausage machine model of the brain.

If V1 is both a sensory map and a motor map, then M1, primary motor cortex, should also play both roles. In fact M1 has been nick-named 'agranular cortex', because it can be identified by a lack of any significant granular-textured layer IV, which is where sensory input enters the other maps. This might mean that it has no sensory inputs at all, but I think it is probably more accurate to say that it has a *reduced* granular layer, which simply implies that its sensory inputs are comparatively sparse. One might expect M1 to receive some kind of

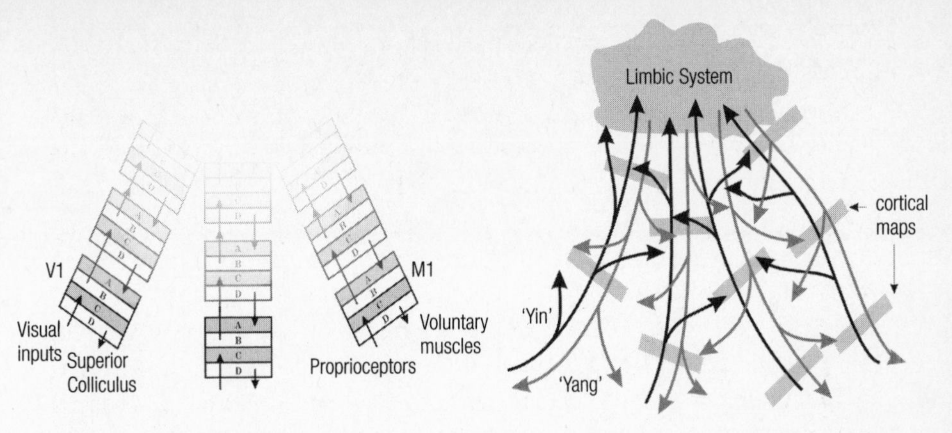

Figure 20 Tripod model

proprioceptive input; signals about the present dispositions of the muscles and joints that it controls. These signals might not be raw sensory data, direct from the muscle receptors; they might be the result of complex processing in another chain of maps, and in fact M1 sits exactly adjacent to the primary somatosensory map, S1, which lies in the parietal lobe and would be the logical source of such information.

At this stage I should probably point out that my simplification of cortical layers v and vi into a single layer D is a bit misleading. In fact, layer vi carries the signals 'backwards' down the chain from one cortical map to another, ending in the thalamus, while the axons of layer v act like a spur, through which motor signals leave the cortex. So in V1, layer v sends motor signals to the superior colliculus, while layer vi continues on, back to the thalamus, joining up with the optic nerve signals that form the start of the other signal chain. V1 is not the only cortical map to send motor output to the superior colliculus: the Frontal Eye Fields in the frontal lobes also connect here, and seem to control more conscious, voluntary eye movements. Indeed many maps have direct motor outputs from layer v, making them into complete sensorimotor units. So the backward ADAD chain doesn't so much *end* in a motor signal, as give rise to motor signals on the way.

Nevertheless, there are at least two sticks in our campfire that have both sensory and motor signals at the bottom end. So what happens at the top of the pyramid, where the sticks meet each other? One possibility would be that there is some kind of master controller sitting here, receiving processed sensory signals from the CBCB pathways, making decisions and then sending down abstract motor commands though the ADAD pathways. But I don't think this is right. The whole idea of a central executive at the top of a hierarchy of command sounds deeply false to me. It's true that the frontal and prefrontal lobes are implicated in long duration planning and our sense of volition, but the frontal lobes are just the stick that has M1 at the bottom; it doesn't fit with the idea of a master controller. A central control module wouldn't explain why the various motor pathways run along the same chains as the sensory pathways either. In fact my guess for what should be at the top of the tripod is the limbic system, a fringe of structures including some of the oldest and simplest parts of cortex and containing the hippocampus, which is implicated in the laying down of new, episodic memories. Why this might be so is something I shall come back to when I talk about consciousness in chapter 18.

This doesn't imply that the limbic system is a master controller, because I don't think there is such a thing. I have been talking about these pairs of signal streams as if they were linear, uninterrupted and unconnected but of course each pair of streams passes through a series of cortical maps, and each map performs processing that undoubtedly combines information from both streams. Moreover, the signals leaving the various maps fan out in both directions, so that any one map receives signals from several others and sends signals to multiple downstream maps. The real structure of cortex is a messy network, only loosely arranged into a set of parallel chains. Processing is occurring at all levels, and the net result of all of this distributed activity feels to us like the unified beliefs, desires and intentions of a single mind. Being of one mind does not imply that all the information passes through a single controlling structure.

Naming the two opposing signal streams in each chain of maps is a bit of a problem. 'Feed-forward' and 'feedback' are too ambiguous, since most people talk about sensory signals passing through the sausage machine in the feed-forward direction, while I prefer to think of the senses as providing feedback to the brain. 'Top-down' and 'bottom-up' are less ambiguous, but charged with emotion. Bottom-up signals are those flowing 'into' the brain, away from the primary sensory inputs, while top-down signals flow out, towards the primary motor maps. The problem with these two phrases is that top-down is often equated in people's minds with centralised control. This is because in hierarchical systems like the armed forces or large corporations, information flowing from the top downwards usually emanates from a central source: the chief executive of a pyramid of command. The concept of centralised control is anathema to many cognitive scientists, and also to me. But top-down signals can also exist in organisations that are not pyramid-shaped, and top-down control is therefore not necessarily a centralised, unified thing.

To avoid all the intellectual baggage associated with the phrase 'top-down', I've named the two different signal directions 'yin' and 'yang'. Yin is the signal that passes from the senses along the CBCB pathway, deeper into the brain, and hence yin is a bottom-up signal. Yang is the signal that passes down the ADAD pathway and flows out to the motor systems, and hence is top-down. Don't worry, I'm not going to start spouting any nonsense involving flows of 'chi' or evidence from Kirlian photography – these are just arbitrary terms. On the other hand, this choice of terminology may carry with it some interesting associations. The idea of a balance of forces, as exemplified by the Chinese concepts of yin and yang, is very different from the linear, sausage machine notion of the brain, which sounds characteristically Western to me. Perhaps my yin and yang streams really do seek to maintain a balance. Perhaps the system tries to minimise the differences between yin and yang at every cortical map. If so, then we are looking at something remarkably reminiscent of the two three-dimensional bar charts I talked about in relation to servos in the previous chapter. Yin would equate with the sensory state of the brain, while yang corresponds to the internal mental state – our desired or anticipated or intended state. On that reckoning, perhaps the yang signal is the key to everything; perhaps this is the source of what I've been calling 'imagination'.

CHAPTER THIRTEEN

Hallucinations

What we normally mean by imagination is our ability voluntarily to project visible, auditory or verbal ideas in our minds; to tell a story that isn't actually happening and may indeed never happen; to rehearse, plan and speculate. However, in this book I'm using it in a much broader sense, to include any circumstance in which the brain forms a short-lived internal representation of the world, if only as a mental reminder of where we put our coffee cup, a prediction that the next note in the tune will be C sharp or an expectation that our muscles will shortly respond in such a way that the soccer ball ends up in the goal. Imagination in this broader sense includes attention, intention and expectation, and it needn't be at all conscious. Why some mental projections are voluntary and conscious while others are not is something I don't yet have a good answer for, although I will try to tackle this question in chapter 18. Nevertheless, I think there is a common thread running between all these things, and the link possibly lies with the yang pathway.

We know from brain scans that when we voluntarily imagine a visual scene, various parts of our visual cortex become active. It seems that when we imagine something, we use precisely the same maps in the cortex that we would use if we were actually experiencing it. If we imagine a sound, we imagine it in one or more of our auditory maps. If we imagine a movement, then we do so using our motor cortex, which is probably why merely visualising a sequence of movements is enough to improve an athlete's actual performance. Both sensory perception and mental imagery (or mentation, to use a less visually specific term) take place in the very same regions of the brain and therefore must, at least partially, share the same circuitry. We clearly don't have a separate and specialised 'imagination engine'.

It's as if we are projecting some information back from a higher cortical map to a lower one, in such a way that the lower map responds almost as if it were really perceiving it. There is a strong link here between visualisation and attention. When we attend to something – perhaps a location in our visual field or an itch on our skin – the evidence suggests that we send out signals that enhance the nerve activity in the appropriate regions of certain cortical maps. Remember that the visual field is mapped out (in a distorted fashion) on to the surface of V1, so if we shift our mental attention smoothly across our visual field, we are essentially casting a searchlight across the surface of one or more maps in the visual cortex, enhancing the neural activity of any cells that were already firing, or perhaps causing them to fire if they were just below the critical threshold. It seems most probable to me that the yang signals are the source of this enhancement.

Notice how this logic then extends perfectly, all the way out to the outside world. In chapter 7 I suggested that controlling our visual attention by making eye movements is comparable to addressing different memory locations in a memory bank. In this case, the 'memory' is not internally stored, but immanent in the physical world itself, and the control we have of our gaze (combined with the narrowness of our cone of acute vision) acts to select one location in this external memory at a time. We also have internal spotlights, at various levels of abstraction, which we call 'attention'. Internally, we cast our spotlights around the surface of a cortical map – a map of the perceived state of the world at some level of perceptual abstraction. The chain of yang fibres that I believe may be responsible for this process projects down from map to map in the visual cortex, eventually reaching V1. But the process need not stop there, because the yang pathway then continues from V1 (although this time from layer v, rather than layer vi) straight to the superior colliculus. Perhaps we can surmise that this is the final stage in the visual attentional system, which no longer scans a spotlight around an internal perceptual map but instead drives the eyes directly around the external scene. As far as the wiring is concerned, this all makes perfect sense.

We can see here the bi-directional and multi-level nature of attention. At the lowest level in the visual system, before the cortex becomes involved, our superior colliculi respond quite automatically to strong sources of motion in the visual scene, by guiding our eyes towards them (see chapter 14). The sensory signal thus only travels through one stage of processing before being reflected straight back the way it came, to cause an eye movement.

But suppose you see something out of the corner of your eye that might represent a dangerous animal. Your brain tells you to freeze, and suppresses this automatic reflex to move your eyes, yet you can still 'look at' the stimulus using your internal visual and auditory attentional spotlights, in order to gather more information. If these new clues suggest it is safe to move, or you can't identify the object without looking at it, your brain might then release its hold on the collicular reflex (or add command signals of its own), causing a visual saccade and a head movement that orients your senses towards the source of danger.

You might rationalise this as a 'decision' to look, but it was the sensory information itself that caused the response. Information passed up the yin pathway and was reflected back down the yang pathway at several levels, causing some action. Sometimes it might reflect back almost immediately, but sometimes it will ripple a long way up the chain of maps before bouncing back down. The further it travels, the more it seems to us like a deliberate choice, rather than a reflex.

This all sounds very reactive and passive, but sometimes the signal might travel down the yang pathway *ahead* of the stimulus. It seems to me that the sensory information on the yin pathway is being processed at each cortical level to provide a *context* for explaining the activity at the previous level. A low-level cortical map might detect the whirling shapes of people, which a higher-level map might then interpret as a dance. Knowing that you are at a dance now provides a context for anticipating what might happen next at the level of simple whirling shapes and also helps you to interpret what the shapes mean, if the available information is incomplete or ambiguous.

Let's examine the second of these functions first: the way in which context fills in for incomplete or ambiguous sensations.

Look at the famous Kanisza triangle illusion (figure 21). You simply can't help seeing white edges where a triangle 'ought' to be, even though you know they are not there. It seems to me that your brain is providing you with several different levels of context when you look at this illusion, and these contextual clues combine with each other to force a diplomatic compromise, in which you are obliged to accept the existence of lines that you can't really see. At a low level, the

Figure 21 The Kanisza triangle illusion

angled corners of the circles suggest the existence of line segments, which your brain then extrapolates and joins up into a triangle. At a higher level, you see how the black 'triangle' has been broken up, and some part of your brain interprets this as evidence that the shape is occluded by another surface. Finally, at a still higher level, the existence of the black triangle makes you more ready to expect to see further triangles than any other shape. All of these things, perceived at various levels in the visual system, set a strong context for believing that the imaginary triangle really exists. Yin and yang battle it out at all levels, resonating with, reinforcing and suppressing each other until they finally come to an uneasy compromise: everyone else will be reasonably happy, just as long as one of the earliest stages of visual processing concedes defeat and accepts the existence of lines that it cannot see.

Recognising the current context might also enable one or more higher maps to make predictions about what a lower map is likely to see next. This may well be the mechanism that I was looking for when I started this quest: the ability to make predictions that enable us to compensate for the often quite long delays between sensation and action.

For example, suppose I show you a sequence of objects and ask you to shout whenever you see an object that could be used as a weapon. You'd probably take quite a time to react to each object, as you figure out what it is and apply the most devious parts of your mind to deciding how to hurt someone with it. Now suppose I do the same thing again but this time ask you to shout whenever you see something green. Almost certainly you'd be able to react much more quickly to this challenge, because determining whether something is green requires much less processing than deciding how harmful it might be. But how exactly does that make your reaction quicker? If the information still has to ripple right through to consciousness and a deliberate decision needs to be made, then I'd expect it to take about the same amount of time, regardless of the task. So if you are able to respond more quickly in the second exercise, surely you must somehow be 'programming' your brain in advance to respond to the question? Once programmed, the shout is presumably triggered directly by the perception of green, and you no longer have to think about it. It's a kind of shortcut – a programmed reflex, in which most of the thinking has been done in advance.

This is a very odd phenomenon in several ways, but it does suggest to me that one of your lower visual maps is being primed to expect the existence of something green, presumably in much the same way that you can choose to attend to one conversation among many, or find it easier to locate church tower symbols on a map once you've seen one of them. It sounds like a consequence of predictive attention.

One of the most interesting ways that we prime our brains to make predictions is when we listen to music. I remember one particular recording of a piano piece, which finished with a glissando that never quite made it to the last note. This failure to satisfy my expectation used to drive me nuts. 'Do re me fa so la te . . . silence!'

The number of tunes we keep stored in our memory always astounds me. Just hearing a couple of notes is often enough for us to recognise the tune. The easiest music to listen to is very predictable and easy to keep ahead of, but as we get more experienced and sophisticated, we prefer music where the logic is harder to fathom; in which the composer leaves out whole chunks for us to fill in. It seems to me that music stimulates us partly because it tests our predictive powers to the limit.

Our auditory cortex is arranged as a tonotopic map, in which different locations represent different sound frequencies. It's not difficult to imagine how the pattern of activity across the surface of this map might trigger the recognition of a tune (or several equally possible tunes) in a higher map. This context could then project predictions back down along the yang pathway, priming notes that might occur next. When one of those notes does happen, it reinforces the link to one particular tune, which causes an even stronger priming signal for the next note in the sequence, and perhaps suppresses other tunes that no longer fit the pattern so well.

The ability to do this with music is not really much of a survival skill, but we do more or less the same thing with spoken language, enabling us to keep just ahead of what someone is saying and thus be ready with a response. As a general principle, almost every part of the brain benefits from such a capacity to predict multiple possibilities and strengthen them into probabilities. Even the flexing of muscles requires prediction, which is why we can make such an ass of ourselves by picking up a box that we thought was heavy, only to discover it is empty.

So, yang signals may be involved in mediating attention and anticipation. It may well be that they are the source of our *in*tentions, too. In my interpretation of the cortical wiring, the yang pathways terminate in (or spawn) various motor outputs to sub-cortical areas. At the terminus of the occipital lobes we have the connections leading from V1 to the superior colliculus (V1, of course, is arranged as a map of the visual field). Likewise, at the terminus of the frontal lobes we have M1, the primary motor map, upon which the joints of the body are mapped out in approximately the right spatial relationships. M1 feeds signals to the voluntary muscles.

It's not entirely clear what actual processing occurs in M1 (or, for that matter, V1) but what is known is that clusters of cells are associated with each joint and that each cell in a cluster represents a specific direction of motion of that joint. Cells representing similar directions are close to each other, while cells representing opposing directions are on opposite sides of the cluster. To visualise this, take half an imaginary orange, place it cut side down on a table and view it from above. Now poke cocktail sticks part-way into the orange at various points, with the sticks pointing towards the centre. Imagine the orange is one cell cluster and the point where each cocktail stick penetrates the skin is a different cell in that cluster. The direction of movement coded for by that cell is represented by the direction in which the cocktail stick points. When we decide to move a limb, a large percentage of the cells in each relevant joint cluster will fire, with the peak activity centred on the direction in which the limb is to move. The actual movement will be very precisely determined by the average (the vector sum) of all the directions represented by these firing cells.

I wonder whether a part of M1's function could be described as *servoing* the limb? If so, then the signals entering M1 from secondary motor maps would equate (in my servo model) to the 'desire' signals, which specify the intended future position of the limb. I haven't yet studied the motor cortex deeply enough to know whether this idea holds water but it is an intriguing thought. Perhaps the secondary map specifies a desired state of the joint via the yang pathway, while proprioceptive signals describing the current actual state of the joint enter the (sparse) layer IV on the yin pathway. Perhaps M1's job is then to work out how much energy to put into the limb in order to minimise the difference between the desired and actual state.

Again, I'm speculating. But if there is any truth in this, then we can extend the principle upwards throughout the chain of frontal lobe maps, in exactly the same way that a chain of servos created higher and higher levels of control in my model glider. The difference is that cortical maps, because of the way they compute things as patterns of activity on a surface, are potentially much more powerful than simple servos.

I'll try to give you some hints about the nature of this computing power in the coming chapters but most of it is too complicated for a book of this type and to be honest I haven't worked all of it out myself yet. For the moment, just focus on the idea of a cortical map as a servo. A servo needs two sources of input and at least one class of output. One input is sensory, and this we can equate with the yin inputs to layer IV. The other, which the servo needs to compare with this sensory signal, is a desire input, which must arrive from a secondary map down the yang pathway, presumably into layer II cells. The yang pathway then continues out of the map through the lower layers V and VI. In the case of M1, layer V contains the large and powerful pyramidal neurons called Betz cells that send outputs to the body's musculature. This accounts for the three pathways needed by a single servo but leaves us with the up-going yin path, emanating from layer III. What might that represent? Well, perhaps it is a completion signal – a measure of the remaining difference between the desired state and the actual state of the servo. If so, then this provides the secondary map with precisely the information it would need to determine how close it is to satisfying its own desired state, as placed on it by an even higher map. In other words, this signal provides the sensory input (actual state) for the secondary map, which itself is a servo.

Bearing in mind the mass of flaky assumptions I've made, such a wiring scheme works out rather nicely. We can imagine a hierarchy of maps in the frontal lobes, through which very abstract motor intentions ('walk forward at a specified pace') give rise to sequences of lower-level intentions ('swing this leg forward by a specified amount'), right down to the movement of individual muscle groups. Each stage behaves rather like a servo, which specifies a lower-level goal and then monitors the achievement of that goal, passing back the current degree of compliance to the servo above. On this basis, the yang pathway is the feed-forward pathway, carrying motor commands down the chain, while the yin pathway is the feedback pathway, telling the maps how close these goals (which would often represent steps in a sequence) are to being met. This fits perfectly with the description of yin and yang for a (predominantly) sensory chain such as the visual system: in a sensory chain, yang represents attention or anticipation, whereas in the motor chain it represents intention; in both cases the yin chain carries sensory impulses which confirm or deny the completion of a part of the intention, or the truth of the anticipation.

If any of this is true, then it seems my tripod model has some interesting qualities. Turning this into a practical system for Lucy is nowhere near as easy as I make it look, but at least I have a terminology and a large-scale wiring diagram that makes some sense and which, furthermore, manages to unify a number of otherwise seemingly disparate properties of the brain. On this basis, attention, intention and expectation are fundamentally similar processes. The only difference between them is that attention and expectation operate along chains of maps that are predominantly sensory in function (like the visual system), while intention applies to predominantly motor chains (such as the chain of motor sequencing 'servos' in the frontal lobes).

The whole network is wired up into a sort of tripod structure, with yin signals rising up each leg, and yang signals cascading down. The yin and yang meet at every cortical map, so that up-coming signals can trigger reflections down the yang pathway, and yang signals can enhance or suppress the signals rising up the yin pathway. This forms a maze of yin-yang loops, in which each signal is affecting the other. The whole system tries to maintain a balance of yin and yang, minimising the differences between the patterns of signals at each level, and hence minimising the differences between the 'desire' and 'actual' signals of what is, perhaps, a network of structures that operate something like servos.

The fundamental difference between ordinary servos and cortical maps lies in the cortex's ability to handle *sequences* of activity, rather than simple linear relationships such as those used to servo an electric motor to a given position. In the case of cortical motor chains these sequences would be used to combine muscle movements into complex repertoires, for example the chain of movements involved in walking. In sensory chains, the sequences enable us to make predictions about cause and effect, or perhaps anticipate the next note in a tune.

So far, this way of looking at things feels very promising and seems to explain a lot of things. Most importantly, it unifies so many apparently disparate functions of cortex, which is exactly what I am looking to do. But there are still huge gaps to fill: what exactly is going on inside each cortical map? How does cortex make predictions and carry out sequences of servo actions? How does all this fit with what is known about V1 of the visual cortex, because none of the existing theories seem to have much to say about servoing, nor do they explain why V1 sends signals to the superior colliculus?

I think it's time to start exploring some of these things in a practical way, by trying to build Lucy a brain. Or at least, a few fragments of a brain.

CHAPTER FOURTEEN

Here's looking at ya

Frogs are very stupid animals. This almost goes without saying. Anyone who has spent half an hour trying to persuade a bunch of frogs that the garden pond which was there last year no longer exists and they'd do well not to try swimming in the new flower beds, will know what I mean. But our own ancestors were amphibians once and at the core of our brains we still carry around, and make use of, neural structures that originally evolved for managing our swampy existence, which have now been sidelined and specialised into something rather less central to our being.

One such structure is called the optic tectum in frogs and the superior colliculus in mammals. As I've already mentioned, the superior colliculus handles what is known as orienting behaviour – moving eyes, head and body to align our senses with a stimulus. If we see a movement out of the corner of our eye, we can barely stop ourselves from glancing at it with our eyes and twisting our neck and body as necessary to bring it into view. Much the same happens if we hear a sudden noise (see chapter 8).

This sort of orienting response is a large part of what it means to be a frog. Frogs aren't visually aware of anything that isn't moving – if it stays still, they simply don't see it. But when something does move, the size, speed, direction and shape of the moving stimulus are all the information a frog needs in order to act. If the moving object is small, it flicks out its tongue towards it and tries to eat it. If it is large, it is assumed to be a predator (or a looming gardener) and the frog leaps away from it and flees towards the nearest shelter. Such feeding and fleeing behaviours are triggered by the optic tectum, which constitutes the very highest part of the frog's visual system. For all I know, if the stimulus is green and moves in a sexily frog-like way, the optic tectum might even be the part of the brain responsible for triggering a third behaviour beginning with 'f', thus completing the frog's rather limited repertoire of choices.

The optic tectum provides the highest level of visual processing in amphibians but the colliculus is a long way from the top of the mammalian visual hierarchy. It still controls reflex orienting responses to visual movement and (via the inferior colliculus) to sudden sounds and thus it retains some functional connection with the visually guided reflexes of frogs. I guess there may even be vestiges of other behavioural reflexes that remain within its remit too, but when higher-level behaviours involving eye movements are called for, these are initiated by more recently evolved brain centres, especially the cerebral cortex. Yet these higher levels of control are not completely separate and parallel systems – they send commands directly to the superior colliculus, suppressing its own reflex movements and replacing them with similar movements driven by more sophisticated sensory cues.

What does the superior colliculus do that makes it the logical stopping-off point for all kinds of visual saccades and other visual or auditory orienting behaviours? It seems that the answer lies in its ability to perform a particular kind of coordinate transform. Every visual stimulus occurs at a particular location on the retina, and therefore is mapped out in retinal coordinates. But to look towards it, we need to know what angle to point our eyes, measured relative to our head.

If the stimulus appears just to the right of centre in our visual field, and our eyes are currently pointing slightly to the left of centre in our heads, we may actually need to look directly ahead in order for the stimulus to fall exactly on our eye-line. Alternatively, if the stimulus is to the right of our visual field and our eyes are already pointing as far right as they will go, we won't be able to focus on the stimulus simply by moving our eyes, so we also need to work out where to point our head, as an angle measured relative to the body.

So, part of the superior colliculus's job is to convert information mapped out in retinal coordinates into different coordinate frames: head-relative and body-relative. It doesn't matter whether the stimulus was a sudden movement (in which case the superior colliculus can perform the task entirely by itself) or something more sophisticated (which requires the involvement of cortical processing). Either way, these coordinate transforms need to be carried out, which must be why various cortical maps such as V1 and the Frontal Eye Fields make use of the superior colliculus for arbitrating and computing eye and head movements.

The fact that the superior colliculus is a common factor in all eye movements, and eye movements are crucial to vision, is reason enough to build such a structure into Lucy's brain. But it turns out that conditional coordinate transforms are far more universal and important than this alone. Even servo action is a closely related concept. So, since the superior colliculus is a convenient place to explore how neural maps can perform such transformations, giving Lucy a superior colliculus seems a natural first step.

Quite a lot is known about the structure and behaviour of the superior colliculus, and there are various theories to explain its operation. I've done my best to take no notice of these theories at all. If I reinvent the wheel, then so be it; but it's important to me that I work these things out for myself, because I'm starting from a somewhat different place from most neuroscientists and doing this project with a different aim in mind. I have tried to absorb the main facts about the superior colliculus's structure, and then explored ways of making the system work that fit with these facts.

One such fact is that the superior colliculus is made from several layers of neurons and each layer forms a map of visual (or other kinds of) space. Another fact is that, just before a visual saccade takes place, a large blob, or dome, of nerve activity builds up across one of these maps, centred on the direction of the stimulus in visual coordinates. In other words, all the neurons in a certain region fire but the ones towards the centre of the region fire most strongly, so plotting a graph of firing rate against position will produce a dome-shaped curve whose highest point is centred over the position of the stimulus in the visual field. Once this dome of activity has risen beyond a certain threshold, another group of cells bursts into action for a short while, initiating the saccade. After the saccade, a smaller cluster of cells seems to hold the eyes in position, keeping them locked on to the target.

This notion of large domes of activity interests me. We see a similar process occurring in the motor cortex and also (assuming that I'm making the right inferences from single-cell recordings, which of course can't tell whether a firing cell belongs to a large group or not) in the parietal lobes and elsewhere. Intuitively, you might think that signals pass through the brain in nice, neat, data streams, like bits

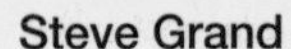

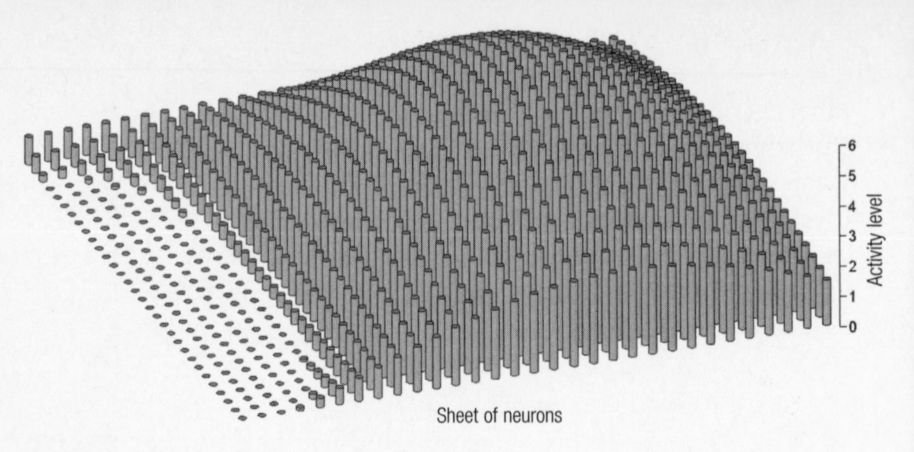

Figure 22 Activity dome in an array of neurons

flowing around the wires in a computer. Those people who work with very small neural networks will be especially biased towards such a conclusion. In reality, wherever you look in the brain, you see vast swathes of cells firing off together, often creating large, diffuse blobs and domes of activity. It turns out that such large-scale patterns, moving around neural surfaces that are mapped out in some kind of spatial or abstract coordinate frame, are a very powerful principle. I would hazard a guess that the right kinds of interaction between such domes of activity might even constitute a general computational scheme. There could be some big secrets locked up in this apparently fuzzy mechanism.

To give you an idea of how important such diffuse blobs might be for making computations on mapped surfaces, consider the task of trying to discover the mid-point between two spatially distinct stimuli. This is a pretty artificial thing to want to do, but imagine an array of neurons, upon which two points, A and B, are represented by a pair of firing input fibres. The task of the map is to produce a third point of activity, midway between the other two. How do we do this? It is easy for us, looking down on the map, but for the neurons to do this by themselves requires them to interact with each other. Perhaps we could try sending out a wave of activity from each point, as if we'd thrown two stones into a pond, and see where the ripples interfere? The trouble with this is that the maximum interference occurs along a line between the two, not at a point, and if the ripples aren't synchronised, this line isn't even in the middle.

Another way to look at this problem is holographically. Suppose each active cell made some contribution to the activity level of every other cell in the map, with the contribution falling off over distance. One stimulus alone would now produce a broad dome of activity. The second stimulus would also produce a dome if it was on its own, but if both points were active simultaneously, their contributions would be added together, and the result would be a single dome, whose peak (or at least, whose centre, depending on the shape of the dome) lies exactly mid-way between the two stimuli.

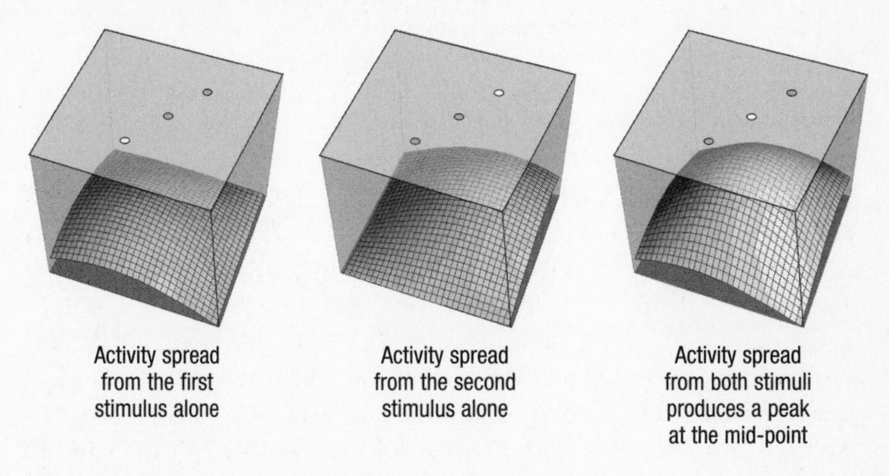

Figure 23 Combining activity domes to find the mid-point

This is a solution to a fairly pointless problem but I hope it demonstrates that input stimuli to a map can interact in interesting ways as long as they are smeared out so that every point on the map receives some contribution from them. Then every data point overlaps with every other, and each column on the map can perform a simple calculation using nothing more than its share of these overlapping input signals, with the result appearing as a new pattern of activity. I call this a holographic approach because it shares important similarities with an optical hologram, in which each point on the original object makes some contribution to every single point on the film. Like holography, which is a powerful computational concept with potential far beyond the production of pretty three-dimensional images, this smearing out (or convolution, see chapter 8) of signals may have implications for many of the brain's functions, including the computing of conditional coordinate transforms.

How might we use this idea of domes of activity to compute the coordinate transforms involved in saccading our eyes towards a visual target? Producing the domes themselves is not difficult for neurons to do. The fastest method is simply for every neuron on the map to receive a connection from every cell on the source map supplying the stimuli, with the effectiveness of each connection proportional to its length. However, this is very expensive, biologically speaking. If the retinal signals arrive at the superior colliculus through a million nerve fibres, then every one of the millions of cells in the first layer of the superior colliculus needs to have a million input connections; one to each input fibre. A slower, but much cheaper method, is for each of the cells in this layer of the superior colliculus to send local connections to a few of its own neighbours and for the total activity of each cell to build up over a period of time. A single input fibre might directly trigger only a single cell in the superior colliculus but this cell will leak some of its activity to its neighbours, causing them to fire. They in turn will leak activity to their own neighbours, and so on. Over time a dome of activity will start to spread and build up. If it is left too long, the network of cells will saturate, until all of them are firing at maximum rate, so something has to monitor the total level of activity and trigger a saccade when the activity level rises past a suitable threshold. This gradual build-up followed by a sudden collapse sounds encouragingly similar to what is known of the real colliculus.

But what of the coordinate transform? Suppose we provide the superior colliculus with two inputs: one of them marks the position in the visual field (in retinal coordinates) where the stimulus occurred; the other marks the direction the eyes are pointing (but this time in head-relative coordinates). Now let these two signals spread out and add together, exactly as we did in the experiment above to find the mid-point between two stimuli. What happens? Does it do the trick?

Sadly no, it's a bit of a disaster. What happens, as we've seen, is that you get a single dome of activity, whose centre lies midway between the centres of the two stimuli. This isn't what we want. Nevertheless, if you get the conditions right, it turns out that this system has an interesting property. Figure 24 shows the result of combining two domes, seen in section. Look at the heights of the dome where it intersects the left and right edges of the right-hand graph. Amazingly, these two heights are more or less proportional to the amount of force we need to apply to the left and right muscles of the eye, in order to bring the gaze to centre on the visual stimulus.

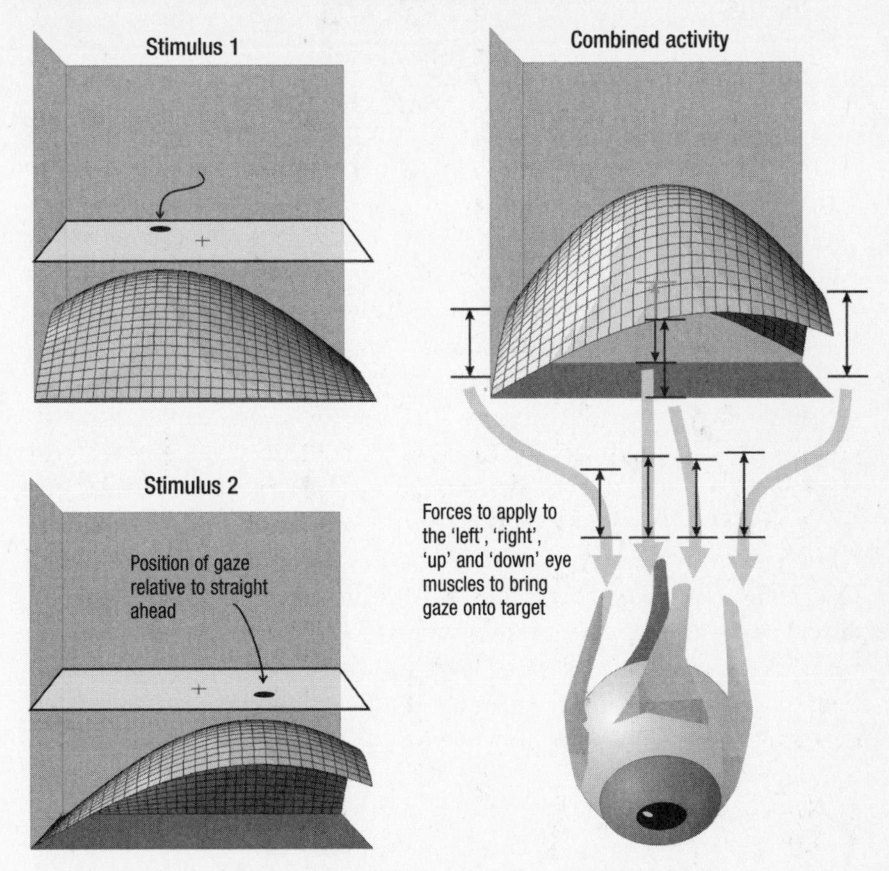

Figure 24 Model of the superior colliculus, using activity domes

If the eyes are looking straight ahead, and the stimulus is in the centre of the visual field, the two domes interact to produce another dome, which is also centred, and hence produces the same levels of activity at both edges of the map. On the other hand, if the stimulus is to the left of the visual field, and the eyes are pointing towards the right of the head, then the two domes are in different places, but their combined result is still centred. If we apply the signals emanating from the two edges of the dome to the eye muscles, they will now pull equally and cause the eyes to centre, bringing the stimulus into the centre of the visual field. This principle continues to work, regardless of the relative positions of the eyes and the visual stimulus. It also works in the up-down direction as well as the left-right one. Simply read off the signals from the centre of each edge (although in practice I do something a bit more sophisticated) and apply these to the up, down, left and right muscles of the eyes. As soon as a movement stimulus (remember that the retina is designed to be highly sensitive to movement) causes a large enough dome of activity to build up in the superior colliculus, it triggers a second dome that is the sum of the stimulus dome and the proprioceptive (eye position) dome. The result drives the eye muscles in such a way that they point directly at the stimulus.

In fact, this doesn't work as well as it seems and I'm glossing over some details but although it's a bit of a fluke it does show the kind of coordinate transformation I'm talking about. Eventually I went on to build a rather different (and in some ways more interesting) mechanism for Lucy, in which the maps learn over time and, through experience, the domes become warped to correct for the odd geometry of her head and eyes. This new mechanism can also produce complete sets of head and eye movements in a single step (and if Lucy could twist her body, it would produce body movements too). The mechanics of this are too complex to explain here, but hey, you aren't building the robot. All I wanted to do in this chapter is give you a flavour of how large blobs of neural activity can be made to perform computations, especially coordinate transforms.

The general structure of Lucy's final superior colliculus (not the one I described here) is shown in Figure 25. Notice how it relies on an input from a higher source (in the simplest case, directly from the retina, but, for more advanced saccades, from the cortex), plus an input from proprioceptors measuring the current direction of the eyes. The target direction in retinal coordinates is therefore a 'desire', and the proprioceptive signal in head coordinates reports the 'actual' state of the eyes. The system reacts in such a way that the actual and desired conditions become superimposed. Does this sound familiar at all?

A modification of this basic idea can be used to construct a servo and before you stopped me from working and made me explain all this stuff to you, I was busy developing a (cortical) system based on these principles that would enable Lucy to control her various arm muscles in order to point to a given direction in body-relative space. The similarities between the two structures are quite striking and even though the superior colliculus is a sub-cortical structure, I feel like I

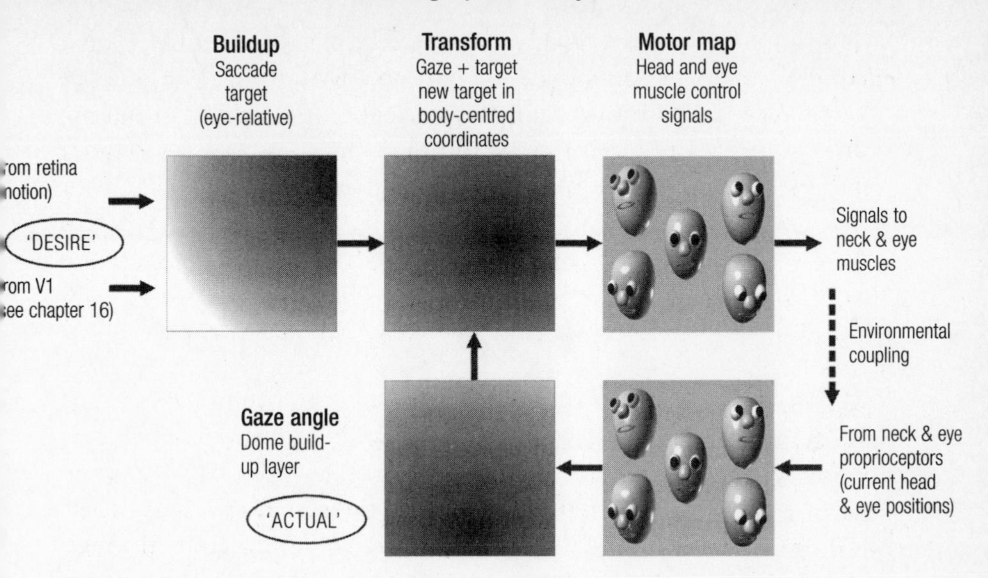

Figure 25 Lucy's superior colliculus (using a slightly different approach from that described in the text)

am beginning to glimpse some of the common themes that make cortex into cortex into cortex, and which thus allow one basic computational mechanism to take on so many roles in the brain.

Now that Lucy has the ability to look in the direction of a visual or auditory stimulus, I need to look more closely at V1, the first cortical stage in processing whatever scene she chooses to look at. To start with I will have to step back from all this talk of servos, because the conventional account of V1 makes no mention of them. First I must create a V1 for Lucy that can do what the experts say our own V1 does. Then I'll take a look at what use this might be.

CHAPTER FIFTEEN

Growing up

I'm sure it's all meant to be about summer days feeding the ducks in the park. It's sharing popcorn and tissues while rooting for Bambi at the cinema. It's proudly attending their first nativity play (our son was a shepherd: 'They aren't real sheep,' he advised the audience in a confidential tone during the Virgin Mary's carefully rehearsed pause for dramatic effect). Peek-a-boo, hide-and-seek and treasure hunts. Tooth fairies, gold stars and graduation day. These are what parenthood is meant to consist of.

In your dreams, sucker! Being a parent is smelly nappies. It's searching for lost homework and having endless rows about why bedroom floors are for walking on, not a convenient alternative to cupboards. When it comes down to it, being a parent is mostly about practical issues like making sure they've got a packed lunch or, in Lucy's case, fully-charged batteries. But then, who said that growing up was easy?

I'm not sure why, but it seems that most people, including many artificial intelligence researchers, think that artificial intelligence will spring forth fully formed. In a laboratory somewhere somebody will one day flick a switch and an android will whirr instantly into action, open the conversation with a remark to the effect that it thinks, therefore it is, then move straight on to solve unified field theory. But when you think about it, even Albert Einstein was over four years old before he learned to tie his shoelaces. We are intelligent because we learn, and learning takes time. Lots of time. 'Intelligence,' as Jean Piaget defined it, 'is what you use when you don't know what to do.' But once you've done it you are able to learn from your experience, lay down new knowledge or a new skill, so that next time you can do it without having to think and can use this new information to help you work out what to do in other novel situations.

Lucy, like the infant Einstein, will have to learn everything from scratch. And I do mean everything. Babies don't seem to do much for their first few weeks of life, but in reality they are learning at a staggering pace. At no other time in their lives will they have so much to learn so quickly, and all without access to someone else's advice. It's simply that the things we learn as babies have become so innate and unquestioned by the time we reach adulthood that we forget they were ever a puzzle to us. Look at it from a neonate's point of view and almost everything we can do now without thinking takes on a whole new significance. Nobody tells a baby that they are separate from the rest of the world, for example. How do we figure out that our hands are part of us while our clothes are not? Or that mummy is another person, just like us but not under our direct control? How do we make any sense at all of the gibberish that assails our ears, or the splatter of colours streaming into our eyes? Even a simple regular shape like a square produces an image that bends and warps in what must at first seem an incomprehensible manner as it falls on different parts of our retinas, but in hardly any time at all we adapt to it so well that we don't notice the distortion and might even find this fact hard to believe.

If knowledge were a measurable commodity, my guess would be that we have absorbed three-quarters of everything we are ever going to learn by the time we are a couple of years old. Making sense of the signals arriving along maybe a hundred million *completely unlabelled* sensory fibres and learning to coordinate six hundred interdependent muscles is an absolutely staggering challenge. After this, getting the hang of long division or learning to drive a car seem like child's play.

Evolution helps, of course. By a series of minor miracles, a child's own instincts and body structure, as well as the intuitive (and therefore evolution-modulated) behaviour of its parents, drives it to learn these amazing feats with a striking level of certainty. We can tell that something profoundly non-accidental is going on because all young children learn the same things in much the same order. Jean Piaget's famed interpretation of these stages of child development has had to undergo considerable revision in recent years, but it is still true that most babies develop in a very characteristic and describable way. I shall have to come back to this topic later, since the relationships between genes and environment and between development and learning, and particularly how these influence the structure of our brains, are an important area to think about. But for the moment I want to examine how learning takes place in the brain at a much finer level of detail, using perhaps the most neurologically studied example: the growth of orientation detectors in V1.

Ever since the nineteen-sixties, when David Hubel and Torsten Weisel first described the ability of individual cells in V1 to respond preferentially to short bars of light with a specific orientation, layer IV of primary visual cortex has been avidly studied. Nobody really knows why V1 does this – it has generally been assumed that finding the edges in a visual scene is a first step in identifying surfaces and then individual objects – but a fair amount is known about the structure of layer IV and, at least roughly, how these orientation detectors come into existence. It therefore provides a good starting point for exploring how the generalised machinery of cortex wires itself up to perform specific tasks.

Layer IV, you will remember, is where the axons from the thalamus, carrying visual signals, enter the cortical map. The visual signals are of various types (see chapter 7) and, at least in creatures like us, with binocular vision, include overlapping signals from both eyes. The thalamic axons end up in the midst of layer IV, where they can synapse on to the pyramidal cells of layer III either directly or indirectly, via two different classes of stellate neuron. Some of these stellate cells are excitatory, and some are inhibitory. The general structure of the wiring seems to be arranged so that cells influence their neighbours over varying distances. Nearby neighbours are excited while more distant ones are inhibited.

There are various theories on how this might lead to the formation of orientation detectors, but I am only qualified to describe how it works in Lucy. To test this mechanism out, I wired up a large array of virtual neurons in the appropriate ways and sat Lucy in front of a computer screen, which displayed moving dark bars on a light background (see plate 6). Luckily Lucy has no boredom circuits yet, so she was willing to play along with this game for hours on end. The columns of neurons start out with smooth, circular receptive fields. In other words, each column in the cortical map receives inputs from a region of visual space much wider than the column itself, with the amount of contribution fading off symmetrically with distance.

Imagine we now project the image of a line on to Lucy's retina and examine the behaviour of a small area of cortical neurons straddled by

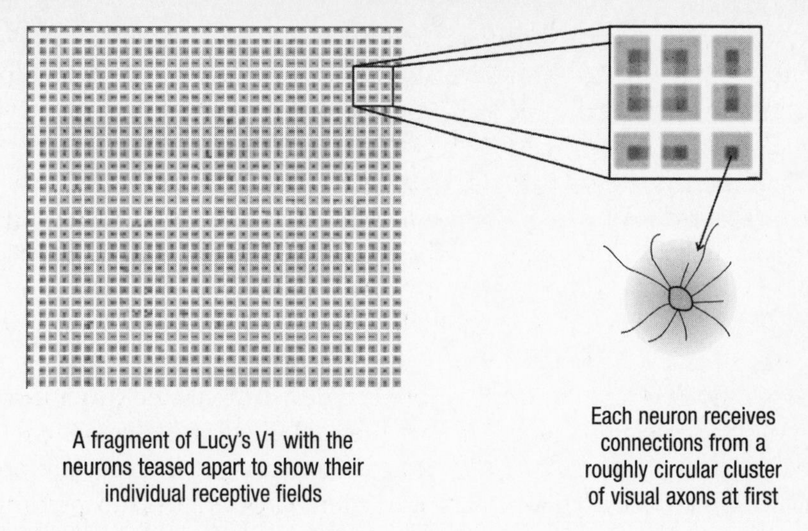

Figure 26 Lucy's V1 before exposure to visual stimuli

the line. Many neurons in this cluster will try to respond to the existence of the stimulus, because they have overlapping receptive fields that touch the line. One or two neurons will respond better than the others, due to their slightly advantageous position. As I said earlier, cells in V1 have recurrent connections that tend to excite adjacent cells and inhibit more distant neighbours. For now, ignore the very local excitation and think about the longer-range inhibitory signal. Every neuron that fires in response to the line will try to suppress its neighbours, and the more strongly it fires, the stronger will be the suppression. The result is a wrestling match between all the interconnected firing cells, in which the cells that fire most strongly will be the best at putting the others down, and hence receive the least amount of suppression themselves. After a short while of this competition, only one or a few cells will remain firing in any given region – an effect called winner-takes-all.

The synapses that make up the receptive field of each cell are programmed with a fairly simple learning rule. This has the effect of making those synapses that contributed to the firing of the cell (by conducting a signal) more excitatory, while the ones that didn't contribute (because they were away from the stimulus) become more inhibitory. The upshot of this is that the cell's receptive field becomes 'tuned' to the signal produced by this particular line. This makes it even more responsive to this particular stimulus in the future, whereas the cells that have been suppressed by lateral inhibition fail to become better tuned and so retain their more general response.

Next time the same stimulus occurs, the newly-tuned cells will fire even more readily than before and hence will win the wrestling match with their neighbours even more easily. Because of this they continue to become better and better tuned to stimuli with that orientation and consequently also less receptive to other orientations. Meanwhile, cells in the local neighbourhood get inhibited by these successfully tuned cells and therefore don't respond to the stimulus, while cells much further away, out of reach of the worst lateral inhibition, are free to tune themselves as they like. So any single long line presented to Lucy's visual field will produce a row of dots of activity in the cortex: each dot is a cell tuned to that orientation, and the dots are spaced out because of the way that lateral inhibition prevents more closely spaced cells from sharing the same preference.

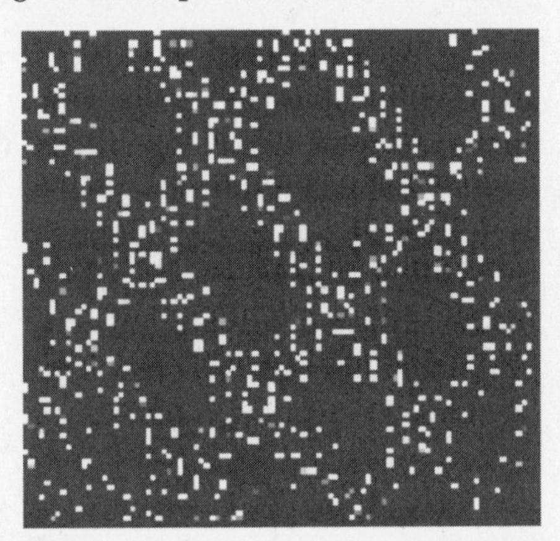

Figure 27 How an image of a checkerboard pattern appears as activity levels in the orientation detectors in Lucy's V1

Now imagine a line at a different angle is presented to the same region of Lucy's vision. The cells that responded well to the first line have become better tuned to that line's orientation, so are not well suited to respond to the new line. But other nearby cells – the ones that lost out in the previous competition – suddenly find themselves with an opportunity. More wrestling takes place and some of these cells win out over their competitors, becoming more tuned to the new angle. Again these winning cells will tend to be spaced out, because cells that are too close together get strongly inhibited. The result of all this, after many trials, is that each cell becomes more and more strongly tuned to respond to a certain orientation and nearby cells will tend to prefer different orientations. Because of the approximately regular spacing produced by the inhibitory range of each cell, the receptive fields will organise themselves into clusters, where each cluster contains an example of every line orientation.

The only thing that is missing from this model is the way that real cortical cells take up different orientation preferences that vary smoothly across space, so that adjacent cells tend to prefer different but closely related angles. The preferred angles rotate regularly as you move across a cluster, starting with north–south and working gradually round to south–north. This is very similar to the smoothly changing clusters of limb movement angles to be found in primary motor cortex.

To achieve this effect, Lucy's V1 cells not only inhibit their more distant neighbours but also excite the cells immediately adjacent to them. They do this in such a way that if one cell is quite well tuned to a new stimulus but not quite well enough to make it fire, some of this pent-up energy gets distributed to its immediate neighbours. If they were non-committal about the new stimulus at first, they may now be encouraged to fire, and hence may find themselves winning the local competition and becoming more tuned to this new orientation. Since one or more of their immediate neighbours provided this extra energy by almost but not quite firing, we know that this new stimulus must have a similar (but not identical) orientation to these neighbours. Consequently, this local 'encouragement' tends to cause neighbouring cells to take up preferences for similar, but not identical (because then they'd end up wrestling with each other) orientations. The result is that cortex arranges itself into clusters (known in the trade as pinwheels – see figure 28), in which nearby cells are sensitive to similar orientations.

Here we have a mechanism through which V1 can self-organise and become able to detect the orientation of line segments. It knows nothing at all about lines to begin with but simply through the mutual repulsion and attraction caused by recurrent inhibition and excitation, it gradually tunes itself to detect and map out the orientations of edges. The result of this self-organisation into clusters of orientation preferences exactly mirrors what we find in real cortex.

To get this to work I trained Lucy by showing her moving pictures

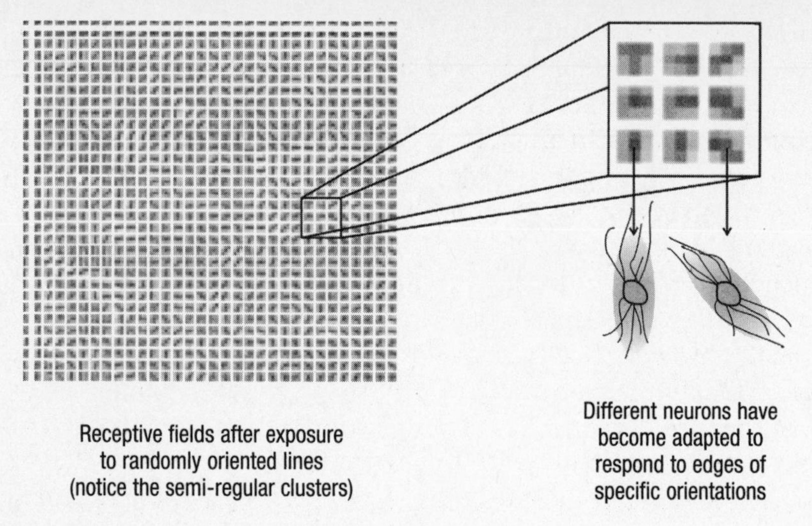

Figure 28 Lucy's V1 after exposure to visual stimuli

of stark black lines at various angles but this was only to make the experiments quick and reasonably repeatable. In fact she can develop exactly the same orientation detectors simply by looking at the real world, just as long as her eyes keep moving so that she is able to sample a wide variety of edges at different locations and orientations. With just a few relatively basic synaptic rules and a combination of inhibitory and excitatory signals, Lucy's V1 wires itself up to do something useful, *simply by being exposed to the world.*

This is part of what I am looking for: a general, unconfigured cortical map that can wire itself up in response to stimuli, in such a way that it learns to perform something useful with the information it is given. Unfortunately, it may well be that V1 in real animals is rather more specialised right from the start than most cortical areas, which makes it hard to extract many general principles from these experiments. On the other hand, similar pinwheel structures exist in motor cortex, and I wouldn't be surprised to find a roughly equivalent set-up in layer IV of the auditory and somatosensory areas too.

However, to describe V1 as 'detecting the orientation of edges' fails to get at the underlying principles of what it is doing. Exactly *why* does it do this, and how can we describe what it is doing in more general terms that aren't tied so tightly to vision?

Perhaps the answer to the second question is that V1 is detecting *correlations* in its input data. The visual world is extremely messy and at the level of individual points of light it doesn't make much sense. If you were to look at the world through three small tubes, so that you could only see three points of light at a time, everything would look completely confusing and random to begin with. But if you looked for long enough you might discover that, every now and then, all three tubes briefly light up with the same colour at the same moment, from which you can conclude that the visual world, even at the three-pixel level, does have some statistical regularity. These particular kinds of regularity are what we call straight lines. Lines and edges produce correlated input signals that stand out from the general randomness. So, at the pixel level, perhaps lines and edges are almost all that the brain can pick out and hence this is precisely what it does. Having done so – having separated the visual world into edges and 'everything else' – perhaps new kinds of correlation can then be discovered that weren't accessible before, such as the fact that edges tend to be connected together, forming boundaries, sides and corners. Perhaps V1 extracts one set of correlations, thus making it possible for other maps to discover new ones?

Yet this is by no means a complete answer to the question of what exactly V1 is *for*. It has always been assumed to be the first stage in the visual processing pipe-line, and that is undoubtedly true. But it still bothers me exactly what it does with all this edge information and why, if it is a purely sensory map, it has motor outputs to the superior colliculus.

Detecting edges is all very well, but there's a lot more I need to understand before Lucy gets a brain that can learn an arbitrary number of things. Nevertheless, vision is absolutely crucial to any understanding of the brain, and certainly for making practical and useful robots. So I'd like to spend one more extremely speculative chapter exploring a specific problem of vision, before finishing up this section with some general principles of brain design.

CHAPTER SIXTEEN

Tilting at windmills

Try this experiment: when I say go, I want you to close your eyes and imagine a scene. When I tried it just now I thought of a Dutch landscape – a wide, flat plain, with a windmill on a bend in a river. It doesn't really matter what you visualise, as long as you do it from a normal perspective, as if you were standing there. Once you have the image firmly in mind, I want you to tilt your head from side to side a few times, smoothly and swiftly rotating your face around a horizontal line passing along your direction of sight.

OK, 'go!'

What did you see? What happened to the landscape? My guess is that it remained anchored to the real horizon, so that what you saw in your imagination was what you would have experienced if you had really been there and tilted your head. If you aren't sure this is what happened, try it again.

This is an easy experiment to do, but I find its implications really quite intriguing. The key point is that the image in your imagination stayed locked to the real horizon as you tilted your head, which surely implies that *your brain had to tilt the imaginary image in the opposite direction* to compensate for the rotation. If it hadn't done this, then the imaginary landscape would have remained locked in the same position relative to your head, rather than relative to the horizontal. This horizon-anchoring effect even works in three dimensions, so you can imagine a scene and then swivel your head around (I mean really swivel it, not just imagine you are swivelling it) in all directions and everything seems to remain perfectly anchored in space.

If you try hard you can force the mental image to remain locked to your eye-line, but when I do this I usually find that the image splits into two, with an effect like a pair of scissors opening. Even though one image tilts with my head, a ghostly version remains stubbornly locked to the horizon. I find that mental images of smaller objects are a little easier to 'bring with me' as I twist. Perhaps the easiest of all is to imagine wearing a pair of thick-rimmed glasses, which of course would normally follow your face as you tilt it. Even then I still see a ghost pair of glasses swing the other way. If I try not to visualise anything at all then swivel my head quickly, I see a strange sort of orange-black separation, as if there was a permanent 'horizon line' in my mental vision that only shows up when it is forced to rotate in order to track the external horizon.

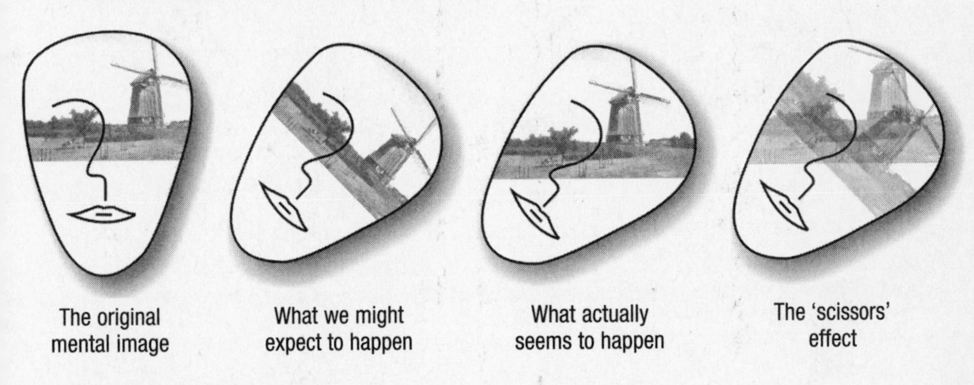

Figure 29 Tilting at windmills

So what does this mean?

To be honest I haven't a clue, but it's odd, don't you think? Are we really rotating an image in our brains, in the same way that we can rotate a digital photograph using image manipulation software? Or are we applying some kind of symbolic reference signal that tells our visual system where 'down' is, and it is this reference alone that is being adjusted? This sounds more respectable as an idea, and the obvious source for such a signal is the balance sensors in our inner ear, but it isn't any kind of an explanation in itself. We can't just pass over it with an airy wave of the hand, because it begs a lot of questions about how such a reference signal is represented and applied to the mental image.

Perhaps it is something between the two extremes. The more I play with my visual imagination, the more it seems like I am able to track the movement of only a small number of key points on my mental image, rather than a detailed picture, and then the picture reconstructs itself when my head is still. In fact, mental images are very much like this anyway: you see the detail when you specifically look for it but only the broad outlines seem to remain permanently present.

Nevertheless, I can do quite sophisticated rotations in my head. For example, I can simultaneously imagine two jigsaw pieces and rotate one until it fits into the other. How could such a *partial* rotation be performed without something genuinely being rotated inside my head? I find that I can easily rotate two mental objects individually, but am completely unable to handle three, so perhaps the answer to this puzzle is that each hemisphere of my brain deals with a different image and since I have only two hemispheres, I can handle only two images. Some support for this idea comes from my personal impression that I can manipulate two imaginary objects side by side more easily than I can if one is above the other (which makes sense in terms of the way that the left and right halves of the visual field are distributed between the two hemispheres).

If this applied only to mental images it might be a little easier to explain. However, a neuropsychologist friend, Peter Halligan, tells me that there is a rare but interesting affliction known as Visual Inversion or Metamorphopsia, in which sufferers (often as a result of stroke) sometimes see the outside world, or parts of it, turn upside-down for a short period. One sufferer even reported that everything in the right half of his visual field looked completely upside-down, while the objects to the left of him seemed only to be rotated forty-five degrees anti-clockwise. Clearly the brain is capable of rotating (or at least *appearing* to rotate, if that means anything different) a real visual image as well as an imaginary one.

Why would it need to be able to do this? It might be connected with the remarkable fact that we are able to recognise an object visually, regardless of which way up it is. This 'visual invariance' is a big mystery, and one of the biggest problems facing anyone working on artificial vision systems. It is relatively easy to program a computer to recognise a given pattern, as long as that pattern is presented at a fixed location, size and orientation, but move it even slightly and the program is likely to fail miserably. We, on the other hand, are able to recognise a familiar object even when it is viewed from a completely unfamiliar angle.

Coordinate transforms and other spatial transformations seem to be a common feature in the cortex, and large parts of the parietal lobes seem to be involved in performing on-the-fly transforms from retinal coordinates into such things as shoulder-centred coordinates, to enable us to reach out an arm towards something we have seen. The direction we need to move a finger changes as we rotate the arm to which it is attached, and this requires a similar rotation of coordinate space to the one we saw when we mentally tilted an imaginary windmill to keep it level with the horizon. One possible connection between such coordinate transforms and visual invariance is the possibility that our brains automatically transform our view of individual objects into 'object-centred coordinates', thus bringing them into a form where they look the same no matter how close they are or which way up they are.

It was while I was thinking about this that I came up with a model for the next stage in Lucy's visual processing that suddenly helped a lot of other things to make sense, or at least apparent sense. I don't feel very happy with this idea as a reflection of reality; indeed, I think it may be a red herring in many ways, but it is interesting and it seems to work. Since it is a good example of the creative, synthetic approach to understanding brain function that I'm advocating in this book, let me take you through the sequence of steps that led me to it. Just bear in mind that there are a large number of inductive leaps involved, most of which I can't yet justify.

The issues of what V1 is actually for and why it should want motor connections to the superior colliculus had been bothering me throughout this project. Then one day I asked myself a silly question that precipitated an interesting line of reasoning. The question was this: why is V1 blobby while V2 is stripy?

When chemically stained in a certain way V1 shows up as a pattern of blobs, surrounded by 'interblobs'. The blobs and the interblobs are known to receive different kinds of signals from the retina and seem to represent either different stages in processing or two quite separate, parallel processing streams. These streams continue to remain separate when the output from the blobs and interblobs of V1 pass into the next cortical map, which is V2. But in V2 the same two data streams take on a stripy form – in fact there are three streams in all, leading to V2's structure of thick stripes, thin stripes and inter-stripes.

Never mind the meaning of these different streams; what interested me was why the blobs in V1 should become stripes in V2. It was only an idle thought, but the more I thought about it, the more odd I found

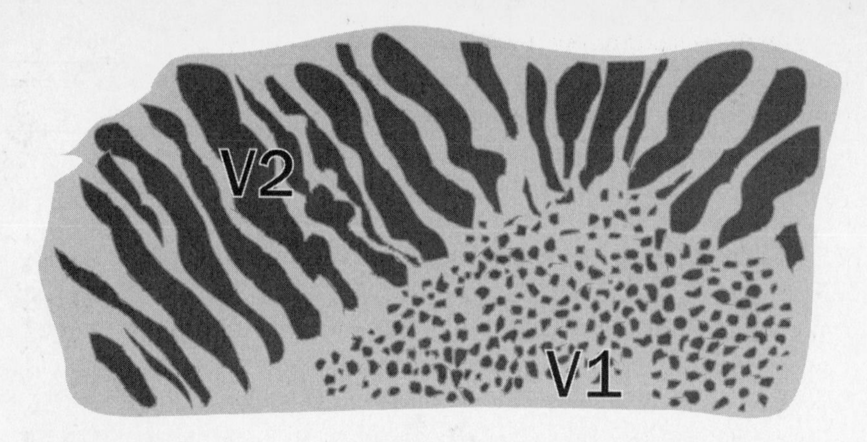

Figure 30 Sketch of cytochrome oxidase staining across V1 and V2

it. When you look at the actual layout of V2 and V1 on the cortex, you find that V2 butts up against one side of V1, with the stripes in V2 pointing away from V1. This possibly makes a bit more sense when you realise that we have two V1s, one in each hemisphere of our brain, each dealing with one half of our visual field. If you were to flatten the cortical surface and fuse these two V1s together to produce a single, roughly circular structure, with the centre of the visual field in the middle of the joined semicircles, you would find that the two V2s almost form a ring around the outside of V1, with the stripes pointing radially outwards from the centre of the visual field.

The fact that V2 lies adjacent to V1 isn't surprising – it's the most efficient way to wire them together. But is there any other significance to the blobby circle surrounded by a ring with radial stripes? Who knows? But there is a mathematical difference between blobs and stripes that is worth looking at: stripes have fewer dimensions than blobs. If you cut a slot in a piece of paper and slide it around over a drawing of some blobs, the pattern showing through the slot will keep changing. On the other hand, slide the slot up and down a set of parallel stripes and the pattern remains the same, so blobs are two-dimensional objects while stripes are essentially only one-dimensional.

Could this mean something? Could it mean that the visual image being processed by V2 has fewer dimensions, compared to that in V1 (whatever that means)? Could it mean, therefore, that V1 'removes' one of these dimensions from the visual signal? This is a pure flight of fancy, but what the heck!

Remember I said that part of the visual system (probably connected with the temporal lobes) might convert the image of an object from retinal coordinates into object-centred coordinates, such that the object 'looks' the same from any angle, and hence is easier to recognise? It's hard to imagine what such an object-centred representation would look like – perhaps it helps to imagine transferring your mental point of view from inside your head to inside the object itself, looking outwards at its surface – but it isn't hard to see how certain spatial characteristics need to be removed from an image of an object in order to get it into such a condition. In fact if we ignore the fact that objects look different when viewed from oblique angles, there are only three such characteristics: the object's position on the retina, its rotation relative to the eye-line and its scale. The same object at different positions, orientations and scales will cast a quite different shadow on the retina and hence will present radically different signals to the cortex. But if you could somehow remove these variables, then an object would appear the same, no matter where it is in the visual field, which way up it is and how far away it is.

We'll come back to V2 and its stripes later, but let's just take a huge inductive leap and suggest that the 'purpose' of V1 is fundamentally to remove the translational (position) dimension from the visual image. In other words, an object (or more accurately a single coloured or textured surface, bounded by a continuous edge) can appear in any part of the visual scene, but suppose V1's job was to move the eyes in order to centre one such surface on the retina? If it could do this precisely enough, it would essentially be removing the translational component from the image, leaving only the rotational and scale components to be dealt with, perhaps by V2.

We know our eyes don't do this precisely, because they are forever darting around from place to place. But suppose V1's job was temporarily to isolate one bounded surface in the image and ensure that it was *approximately* centred, or at least that the eye was fixated somewhere inside the boundary. When you look at plots of eye movements, this is what the eyes tend to do: they enter a bounded surface and then dart around inside that surface for a while, before crossing the boundary and doing the same to another surface. Certainly such boundary crossing is important for working out which surfaces enclose or partially occlude other surfaces.

This is rather heretical stuff, I warn you. V1 has long been viewed as the first stage in a *passive* sensory processing pipe-line, acting rather like a personnel officer sifting through a large number of job applications before sending the good ones on to management for further assessment. What I'm very tentatively and nervously suggesting here is that V1 performs a *motor function*, isolating bounded surfaces from their background and then steering the eyes (using the hitherto completely mysterious connections to the superior colliculus) so as to modify the position of the incoming visual signals.

In Lucy's equivalent of V1 (see plate 6) the cells in layer IV identify edges in the tried and trusted manner, but then other layers of cells in the same map use long-range mutual excitation to accentuate only those edges that happen to join up with each other into a complete boundary (if I'm lucky). This has the effect of isolating one particular boundary for scrutiny, which then causes the superior colliculus to trigger a visual saccade that will bring the fovea somewhere inside this boundary. There is some neurological evidence to support the existence of these additional stages of processing in V1, but here is not the place to describe it in detail. In all probability I'm talking complete rubbish, but never mind – I hope you still find the train of thought interesting.

Let's say the eventual output from V1 is guaranteed to contain a single, strongly enhanced, bounded surface and the centre of the visual field is guaranteed, because of the consequent eye movement, to lie somewhere inside that boundary. Essentially we have now removed most of the translational component from the visual signal (as well as everything other than a single surface from the image). What next?

This is where the hint from the radial stripes comes in. Imagine for the moment that V2 is circular, and that somehow all its cells have wired themselves up to their immediate neighbours in radial lines. The next chapter deals with how such structured kinds of wiring might arise in general terms throughout the cortex, so for now just take it as read. If we trigger any single cell in our hypothetical V2, it will fire with a short burst of activity, triggering its immediate neighbours and starting off two cascades of nerve activity. Two pulses of signals will travel across the map: one directly outwards along the radial connections and one inwards towards the centre. Similarly, if we were to project a precisely centred, circular stimulus on to the map, two circular waves of activity would arise, one travelling outwards and one inwards (like the ripples you'd get if you dropped a hoop into water). Note that all of the ingoing signals would meet in the centre at exactly the same moment.

I hope you're following this so far. I don't really want to get bogged down in technicalities – I just want to take you through my thought process so that you can see how far a silly idea about blobs and stripes can lead and how radical it can make one's interpretation of the brain.

If a circle produces an ingoing ripple that hits the centre at precisely the same moment all round, then what would a square do? The middles of the sides would hit the centre first, while the four corners would get there last. If you plotted the amount of nerve energy hitting the centre over time, you'd get a pattern that is fairly characteristic for squares and rather different from the sudden blip produced by a circle. But here's the thing: it doesn't matter how big the square is, because the ripple will just take longer to reach the centre for a bigger square. Also, it doesn't matter which way up the square is, because everything has collapsed down to a point by the time the activity reaches the centre and points don't have a right way up. The same pattern of activity will occur at the central neuron regardless of the size and orientation of the shape.

So, our hypothetical V1 has removed the translational variable (or at least minimised it, since the pattern still varies somewhat as the eyes bounce around inside the shape) and now the radial lines of our equally hypothetical V2 remove both the scale and rotational variables, leaving us with a (relatively) unique 'essence of shape', which I suppose can be described as the shape drawn in object-centred coordinates. Neat, huh?

And it gets even neater: if all the signals meet at the centre with a characteristic distribution in time, then they must logically pass through all the radii with the same characteristic pattern. So with a bit of extra wiring that I won't trouble you with, plus another layer of neurons, we can ensure that all the cells along a given circumference combine their signals together in such a way that every cell on this second layer receives the same pattern of impulses. In fact we can then dispense with the centre of the map completely, making this model of V2 fit with the ring-like structure that I suggested might be the case in reality.

One more step and we have a complete system. Every cell on this new second layer now receives exactly the same 'time signature' from the shape, but neighbouring cells get the signature at slightly different

Figure 31 'V2' activity patterns for various apple and banana shapes, presented at different positions, orientations and scales

times. That means we could take a snapshot at any single moment and find that the time signature is represented as a travelling spatial pattern on the surface of the map. It's a relatively simple matter to add another layer of neurons (or another map entirely), whose synapses learn to recognise and organise these patterns in almost exactly the same way that the cells in layer IV of V1 competed and collaborated with each other to recognise and organise the different orientations of lines. What we end up with, if you've followed the logic so far, is a map that can learn to recognise and classify different shapes or characteristics of shapes (such as corners and curves), and can recognise those shapes no matter where, how far away or which way up they are.

Hey presto! Of course there are a huge number of qualifications, problems and irritating details that I haven't bothered to tell you about. As you can see from figure 31, the pattern of nerve activity at any one moment is still pretty ambiguous, since the pattern changes quite a lot as the eye moves around inside the boundary. Also, radically different shapes can produce annoyingly similar patterns. The full solution has more to do with the way different patterns are related to each other, how they change over time and how the history of recent patterns can suggest new places for the eyes to look. Also, this system isn't much use for recognising complex, multi-part, three-dimensional objects without still further maps that I don't yet know how to build. I'm not pretending that this is any kind of an answer, but it does seem to work after a fashion, and it's weird and unconventional enough to suggest that it might lead somewhere eventually, so I'm currently using Lucy to experiment with the ideas.

Whether the idea has anything to say about real brains I don't know, but I think it illustrates several interesting things. First, it makes some sense of V1's motor connections to the superior colliculus, and therefore perhaps has something to say about visual attention (which may play a part in selecting which bounded surface is to be examined at any moment and where inside a given boundary the eyes should look next). Second, the idea of V1 using motor action as a means to perform sensory processing seems like a radical shift of viewpoint (and the idea can be taken further, for example to give V1 a role in binocular vision). Third, it offers a vaguely plausible role for V2 and later cortical maps (although I have no evidence whatsoever to support this). Fourth, it offers an object-centred coordinate model for thinking about the problem of visual invariance, which in turn has implications for other brain functions. And finally, it shows how carried away I can get when I ask myself dumb questions about blobs and stripes!

If I combine Lucy's model of the superior colliculus with the cortical models of V1 and V2, and link these to a servo-inspired model of primary motor cortex, I have a complete loop. At the time of writing I have tested all the ideas individually but haven't put them together yet (I've only got one pair of hands, and there are some other things I want to do first). By the time you read this I hope to have got much further, but only time will tell. The acid test will be when I can hold up an apple and a banana in each hand, at arbitrary distances and orientations and find that Lucy has learned how to recognise and point at the banana. This might not seem much of an achievement to you, but it'll make me a very proud father indeed, and choosing between bananas and apples is quite a big deal if you're a robot orang-utan!

CHAPTER SEVENTEEN
The dreamtime

'Do androids dream of electric sheep?' wondered Philip K. Dick. More to the point, though, do hibernating squirrels dream of nuts?

Dreams are so important to us that I can hardly bring myself to ask *why* they are important, for fear of shattering something precious. But this is a challenge I face time and time again in this project: to step from the visionary world of poetry into the practical world of prose without losing my way back.

It's the age-old problem of art versus science, analysis versus synthesis, heart versus head. Why these things are so often considered to be in isolated opposition I really don't know, but it is something we urgently need to address as a culture. An important part of what I'm trying to show in this book is that analysis without synthesis can be a serious mistake, while synthesis without analysis is either impossible or worthless. A mind is the 'more' that is more than the sum of its parts. This is because minds are a consequence, not only of the properties of those parts, but also of their organisation. Analysis helps you to understand the properties of the parts and the individual elements of their organisation, but the qualities of the whole can only be seen when those parts are returned into their proper and complete juxtaposition. Only by synthesis can we appreciate and understand the 'more', but we also need analysis to tell us what building blocks to use, and prevent us from fooling ourselves with mere beliefs.

Creative synthesis is art; diligent analysis is science. To have the one without the other is futile, to keep the one isolated from the other is dangerous and unbalanced and to confuse the two is obstructive. Beauty and truth are like the faintest of stars: to see them you must avoid looking directly at them. You must approach their presence circuitously and reverently, avoiding the monarch's eye. But to do this you need to know precisely and unambiguously where not to look. This is why we need both art and science, poetry and prose.

Dreams must have a biological purpose and knowing this purpose need not diminish them in any way. Indeed, the alternative is for dreams to have no biological purpose, which makes them mere epiphenomena – pointless accidents. For some people, dreams are very important. They see things in their sleep and are then guided by those thoughts through the day. Being of a slightly more prosaic disposition myself, my dreams aren't much help, but I value the short period between sleep and wakefulness for much the same reason. Almost all the creative thoughts that enter my head arrive during this period and it is the most important part of my day. For me, the art of thinking lies in capturing those dreamy ideas before they float away and then converting them to prose that I can store and make use of when my mind has returned to its prosaic but relatively clear-thinking and practically-minded condition. (Reading the chapters of this book, you'll discern this transition in action – the first paragraph or so usually occurs to me as I'm waking up and then the tedious but necessary practical details follow on later.) Whatever dreams mean to us personally, they must first have meant something of value to evolution, or they wouldn't exist. The evidence suggests that all mammals (with the probable exception of the monotremes – the most primitive group, which includes the duck-billed platypus) undergo Rapid Eye Movement sleep, and hence might be said to be dreaming. So why do we all do this? It can't be for any reason that applies only to humans.

Like just about everything else in this book, I really don't know the answer to this question but I have some thoughts that might have a bearing on the subject, which I'd like to share with you. First, let me return to the theme I have been building all along: the search for the principles of self-organisation that are able to turn a generalised, unconfigured cerebral cortex into a series of specialised machines that go on to perform the various sub-tasks of the mind. In other words, my search for the magic that allows a brain to bootstrap itself into a person.

Writing a mind on to the blank pages of the brain is a process that takes many years to complete. Ideally, from a human perspective, it takes a lifetime. I'm sure I am more conscious and alive today than I was when I was six months old; I suspect I am slightly more conscious than I was when I was in my teens. With luck and a good deal of effort,

I aim to be more conscious still by the time awareness is finally snatched from me, although toil and tears, and perhaps senile dementia, may decompose me sooner than I'd like. Most of this organising power, the force that constructs us, is painstakingly supplied by our sensory environment: by the world in which we live. Genes provide only the framework, the impetus and some of the biases.

As I hope I showed in the previous two or three chapters, at the neural level a few quite simple rules, such as the mutual repulsion and attraction generated when short-range excitation is coupled with long-range inhibition, are enough to enable cortical tissue to extract real meaning from the environment. In the case of layer IV of primary visual cortex, this meaning is very primitive and shows itself merely as an understanding that the visual world can be divided up into surfaces bounded by edges. But as soon as these most fundamental discoveries about the world have been identified and learned, more complex and subtle revelations can occur.

Once we know that the world is made from bounded surfaces, we can discover that those surfaces have a constancy of existence, despite the way their shape changes as we view them from different angles. From this we learn that the world is broken up into objects that can be moved around independently and coherently. We then discover that some of these objects have an effect on us and us on them. A major breakthrough occurs on the day we learn that one object can have an effect on another object and the consequence of the conjunction can have an effect on us. For many of us, this day arrives the first time we discover how to make a noise by banging one toy with another; we then go on to explore the boundaries of this discovery in the sand-pit and at the dinner table. Eventually, like our hominid ancestors before us, we learn how to alter an object so that it has new properties, which in turn allows us to manipulate other objects in new ways: we discover how to make tools.

Meanwhile, there are other chains of inference and stepping-stones to discovery. We learn that whenever we twitch our mouths and lungs in certain ways, we hear a sound. We discover that different kinds of twitch make different sounds and that we can copy sounds that we hear coming from other places, such as the mouths of our parents. Pleasingly, we find that our own sounds seem to have an effect on other people, while we discover with less pleasure that certain sounds from other people are sometimes correlated with an unpleasant experience, such as a slap on the wrist, although we soon learn how to stop the bad sounds and invoke the good ones. Ultimately, these tiny revelations about how sonic and other events in the world are correlated at higher and higher levels lead us to discover literature, oratory and companionship. From such tiny acorns, the mighty oaks of the human mind grow.

For any of this to be possible, we need the right infrastructure. Brains don't start out as a random mush. Somehow, the basic elements of the nervous system need to be linked up in the right ways to provide a canvas upon which the outside world can paint a thinking machine. Building this infrastructure is the responsibility of our genes, but genes can work in surprisingly subtle ways.

An entire human being can be specified using a blueprint of around thirty thousand genes. The human brain meanwhile consists of a hundred thousand million neurons, with a staggering one thousand million million interconnections. It doesn't take a genius to work out that there's a bit of a discrepancy here. In many evolutionary neural network experiments (see chapter 3), each neural connection is specified by a different gene, which is clearly not the case in humans. The brain is highly repetitive, of course: cerebral cortex is made from a basic six-layer sandwich and the connections between different cell types in this sandwich show a consistent pattern throughout. So as long as genes are able to say 'make the following thing a million times', we need far fewer genes than the number of connections and neurons might suggest. Even so, the brain is very complex, considering the small number of genes devoted to it. Wiring up the basic structure via explicit rules ('fold tab A and insert into slot B') is not the way things are done. In practice, genes *influence* growing neurons in ways that *encourage* them to link up into complex circuits. The brain is nudged and cajoled into shape, and genes exploit various properties of the physical world, such as the way that diffusing chemicals form concentration gradients, to leverage their own relatively small information content into something much more complex.

One way in which genes are known to exploit the laws of cause and effect so as to construct complexity out of simplicity can be seen in – you guessed it – area V1 of the cerebral cortex. Something I haven't previously told you about V1 is that the signals from both eyes converge on to the cortical surface (each half of the visual field entering different hemispheres of the brain) in a highly organised way. Where the axons initially enter layer IV, the fibres emanating from the two eyes are completely overlapped, like two packs of cards that have been riffled together into one. The cells in layer IV that connect directly to these fibres gather together signals from many adjacent axons but, surprisingly, only select those from a single eye. The cells are monocular, even though the input axons from the two eyes are intimately mingled. Moreover, cells with a preference for a certain eye are organised into groups, forming fine stripes of alternating eye preference as we move across the cortex. The outputs from these cells are then gathered together by further cells in layer IV, only this time the cells are binocular, accepting inputs from both types of monocular cells. Near the middle of a stripe, these binocular cells can only 'see' inputs from that stripe and hence get most or all of their input from one eye. Near the edge of a stripe, on the other hand, the cells are able to gather inputs roughly equally from both eyes and in intermediate locations the proportion of signals from each eye varies smoothly between these two extremes.

Such an intricate and complex structure looks distinctly like it was hard-wired before birth by multiple genes telling the groups of neurons explicitly where to connect, but in fact the mechanism by which it develops (as uncovered by M. Stryker and S. Strickland) is far more ingenious and subtle. Imagine a foetus developing in the womb. The overlapped bundles of fibres from the two eyes have already grown out from the thalamus to reach V1 but none of the cells in layer IV has yet formed connections to them. Now imagine that some of the cells in layer IV are doing their best to grow dendrites out to meet and synapse with any axons that happen to be carrying electrical signals. They'll grow towards anything that is firing but the eventual strength of their connections to other neurons depends on a rule known as the Hebbian growth rule (after the neuroscientist Donald Hebb). This rule says that when input signals are closely correlated in time (hence adding up to a signal large enough to make the cell fire), the specific synapses that conducted them tend to become strengthened.

Now let's move to the developing child's retina and imagine that the neurons there have a tendency (in the absence of anything to look at in the darkness of the womb) to fire off spontaneously. If one retinal cell fires, it is likely to trigger off its neighbours, so what tends to happen is that a spreading wave of activity arises from a random spot and grows across the retina like a bush fire. Then there is a brief lull while the neurons recharge, before another spontaneous burst causes another wave, emanating from a different centre.

Because this spontaneous activity forms expanding wavefronts, any two adjacent cells in the retina are quite likely to fire at approximately the same moment, as a wave passes over them both. On the other hand, the wave travelling across one retina is not at all correlated with the wave on the other and so fibres emanating from approximately the same spot on different retinas are likely to fire at completely different moments.

What do the cells in layer IV do when this activity reaches the cortex? According to the Hebb rule, they have a tendency to grow connections to sources of input that fire together, which means they will grow towards adjacent axons from the same retina, while ignoring the uncorrelated activity arriving from the other retina. It takes a little more finesse than this, including some sharing of information between adjacent cells, but I hope you can see roughly how this purely spontaneous activity in the two retinas leads to the growth of monocular cells arranged into bands from different eyes.

Now the baby is born and its eyes are suddenly filled with new sights. The spontaneous activity dies down behind the barrage of signals from the external world. But this time, these externally stimulated signals are for the most part pretty randomly distributed across the visual field, and not spatially correlated in the way the waves were. On the other hand, because both eyes are now seeing the same image, identical spots on each retina are very likely to be receiving the same signal at the same time and hence will tend to show the same pattern of firing. Meanwhile, back in the cortex, the second level of layer IV cells is busy looking for some activity to connect to. And just like those in the first level, these cells will prefer to gather signals from sources that are correlated in time. This time the correlations occur between equivalent points on both retinas, rather than adjacent points on one retina, and hence the second level of cells tends to grow binocular connections to the first level. From these simple facts we end up with the complex, stripy structure of monocular and binocular cells found in cortex. QED.

What a clever trick. All you need to do is to make sure that the developing retina produces spontaneous activity during the period when the first layer of cells is growing (neurons automatically do this in the absence of any other source of input signals anyway) and then make sure that real visual stimuli are available by the time the second phase of growth occurs. You hardly need any genetic instructions at all to ensure that these conditions are met. Out of this developmental process you then get a complex pattern of monocular and binocular cells, presumably with some useful part to play in the behaviour of V1. If my hunch that V1 is there to stimulate visual saccades into bounded surfaces holds any water, then we can presume that these saccade commands will need to include information about the angle of convergence required to bring the two eyes into register when the surfaces are at different distances. In which case, V1 is the very place that we would expect to find such binocular processing circuitry being required. If Lucy had more than one eye, bless her, I might have been able to explore this idea more thoroughly by now.

What really interests me, though, is how general this principle might be. I'm intrigued by this use of spontaneous but structured nerve activity to form a sort of 'test card' pattern that informs one part of the brain about the structure of another part. I also find it interesting that this initial framework is then modified by real sensory data, which produces new kinds of correlation for the system to learn. Does this constitute a central principle in how the brain wires itself up? Is it a major force for self-organisation? Do such test cards set up the initial framework for the cortical wiring, upon which experience superimposes a more detailed structure, which in turn generates the model of the world through which we act?

One theme that seems to be recurring throughout these investigations into the cerebral cortex is the concept of coordinate frames and coordinate transforms. V1 is arranged as a map of the retina; M1 is arranged as a map of the body's musculature; S1 is a map of the tactile sensors on its skin; A1, the primary auditory cortex, is a map of sound frequencies, presumably organised in such a way that the same notes in different octaves are related to each other. Many maps in the parietal and frontal lobes seem to be involved in making on-the-fly transformations between information in one coordinate space and that in another (as does the superior colliculus, of course). Even our ability to recognise objects might plausibly result from a transformation from visual coordinates into object-centred coordinates. Coordinate transforms are everywhere, and it seems a safe bet that every cortical map represents a particular kind of space, whether it is a physical space, like the visual field or the space that an arm can reach into, or a notional space (more on which in chapter 18).

We can make a bold assertion about these coordinate frames: if anything that happens to us visually is to have an ultimate consequence in terms of our behaviour, there must be a path that somehow transforms the information from visual coordinates into muscular coordinates. The signals that flow from our eyes, through our cortex, and out to our muscles, regardless of the spreading out and manipulation that occurs to them *en route*, must go through some kind of transformation that leads seamlessly from the input space to the output space. It is almost as if a picture of a dangerous animal is somehow being morphed in stages into a picture of a person running, which has the direct effect of firing off the right muscles to cause us to flee.

Moreover, we can assert that when information streams arising from two completely different sensory modalities (for example vision and hearing) converge on a single cortical map (as they presumably must if they are to be compared and used together) then they must by that time be in the same coordinate frame. If we are trying to work out which of the several people in front of us said something, for example, we need to combine what we can hear with what we see, and correlate the two. To do this we need to take the information from our ears (after positional analysis, as in chapter 8) and combine it with that from our eyes; hence the two streams need to be converted into a common coordinate frame.

How might such a complex of morphs arise? The end points in the various chains are fixed by the geography of our bodies. The arrangement of nerve fibres from the eye determines the mapping of V1 and the arrangement of nerve fibres from the muscles in our body determines the mapping of M1. However, I am suggesting that in between these fixed endpoints, we have a whole series of maps that represent intermediate coordinate frames of one type or another. How might these wire themselves up, so that the signals entering and leaving them make sense in terms of the maps to either side?

My speculation is that this might be achieved by using test card signals, flowing into the various cortical inputs. If spontaneous signals with a given spatial structure enter a map and then get twisted around and modified, departing from the map in a new arrangement, then this new pattern carries information about the structure of the map it has just passed through and can make this information available to the next map downstream. I don't have space or enough confidence in the details to go into this much more fully here, but it is a very tantalising idea. Remember the yin and yang pathways? If information passes along the yin pathway from one map to another, it must undergo a coordinate transform at every step. If the yang pathway carries signals in the other direction, then this must undergo the converse set of transformations, so that it ends up in the right format. I strongly suspect that a tension between these two opposing signal streams would be enough to teach all the intervening maps how to wire themselves up into intermediate coordinate frames that interface neatly with their upstream and downstream maps.

If the fibres leaving V1 carried a test card signal, which passed from map to map and ended in M1 (to simplify things for the sake of understanding) then these signals alone would be unable to discover what the destination coordinate frame is until they arrive at the last map, by which time it is too late. All the maps will be in visual coordinates until the last one, when a sudden (and presumably unassailable) jump has to be made from visual space to muscle space. But if yin carries test cards in one direction, starting in visual coordinates, and yang carries test cards in the reverse direction, starting in muscle (proprioceptive) coordinates, then each map has an opportunity to compare these and work out a coordinate frame that best suits both the upstream and downstream maps. Motor cortex would try to make every map use motor coordinates, while visual cortex would try to make them all use visual coordinates, but each map receives a different proportion of coercion from each side, and so will find its own level. My early investigations suggest that the system will ultimately come to wire itself up in an optimal way.

The result should be a set of mappings from the source to the destination (and vice versa) that minimises the complexity of the transformation involved at each step and hence (it seems to me) maximises the chance of each map being able to find useful correlations in its input data. Live environmental data can then come along and find a sensible home in which to live – a place where it can not only find correlations within its own data stream, but also correlate these with information arriving from the opposite direction, along the yang circuit.

This concept is so interesting and has so many ramifications that it could form the basis of a whole book. I can only skim the surface of it here, and hope that you will find it an appealing idea that the large-scale structure of the cerebral cortex might plausibly result from a resonance or tension between yin and yang signals. This is very similar to the hypothesis I described in chapter 13, in which a yin/yang tension between the brain's sensory state and its 'mental' (attention, intention or prediction) state might give rise to servo actions that control our behaviour, only in this case the tension arises as a result of pre-natal test cards generated internally by the body's sensory systems. I'll leave you to sleep on the idea and see whether it resonates with you.

Which brings me back to the question of dreams. I've been suggesting that the brain is in a perpetual dynamic tension that drives its neurons to make, and constantly remake, the right kinds of connections. I'm also offering up the idea that this tension arises from two kinds of influence: the highly structured, spontaneous test card data that forms the initial infrastructure, and the subtle, complex sensory experiences through which this infrastructure is able to bootstrap itself into an active model of the outside world, capable of changing that world to the creature's own advantage.

We know that cortex is in a state of dynamic and very fluid tension, because if you suffer damage to a portion of your retina (as I have), the cells in V1 very quickly reorganise themselves to cover up the gap. I have a scotoma in my left eye, which means I'm blind in a small portion of my visual field. I only notice it if I deliberately move a small object into this region and note that it disappears; normally the world looks seamless and complete to me. Lesion studies have shown that this remapping to cover a flaw can begin within seconds of the damage occurring – as soon as the brain notices that it is no longer receiving sensory data from this region. This shows very clearly that the cortex is a machine in dynamic tension. It is constantly trying to re-wire itself but is held in one particular configuration by forces that only give themselves away when they disappear.

If the cortex is perpetually trying to reorganise itself, what happens to a squirrel's mind when it falls asleep for a whole winter? Squirrels evolved the ability to hibernate in order to slow down their metabolic activity during a period when food is not available. Likewise, we evolved the ability to sleep in order to keep us safe and energy-efficient during the period when our own food is impossible to see and nocturnal predators hard to detect. If a squirrel isn't being reminded about the location of stored nuts and the properties of trees and predators by repeatedly experiencing them, why doesn't it forget all about them by the spring, in the same way that my visual cortex quickly forgot about part of my visual field? Neurons don't hold their connections for very long unless they are refreshed by new nerve activity, so memories need to be rejuvenated. Perhaps this is why mammals dream? Perhaps dreams are, at least in part, our memories – our cortical wiring – recycling itself, to keep the fine-grained structure of our brains from being wiped clean.

We don't dream all the time. In between periods of dream sleep we go through what's called slow wave sleep, during which the cortical cells have a distinct tendency to fire in rhythmic synchrony. Is this the other half of the story? Is slow wave sleep a test card phase (generated by the thalamus), which maintains the underlying infrastructure that otherwise would be degraded by the passage of time, and coincidentally cleans up some of the environment-stimulated rewiring that hasn't proved its usefulness? If a disorganised mass of neurons can be persuaded to wire itself up into a neat structure by the use of test cards, and then decorate itself with a finely detailed ornament created by the interaction of this infrastructure with the correlations inherent in sensory data, then in the absence of these two organising forces, I'd expect the brain to return to being a random mush. Perhaps this is why we dream, and why we alternate dreaming with slow wave activity, no matter what extra intellectual inspiration and imaginative enjoyment our dreams may bring us as a bonus.

This is all absurdly speculative. I can't back up any of my assertions conclusively. They are simply ideas that I think are worth following up, using Lucy to test them. It may take a few years before I realise that I'm barking up the wrong tree and these ideas simply don't work, but none the less I think it is worth sharing them with you. Just possibly some of these hypotheses might be true, and will take us a little nearer to understanding ourselves, as well as helping us to create complex, self-organising, intelligent machines.

I'm probably just fooling myself, but hey, a guy can dream, can't he?

CHAPTER EIGHTEEN

Fusion (and some confusion)

One of the best ways to irritate an academic is to tell them that their brand spanking new 'I'll get a sabbatical year in the Seychelles out of this for sure' theory was actually first proposed by Lock, Stock and Barrel as long ago as 1924 and subsequently shown to be false.

Another really good way to wind them up is to go around calling your robot 'she' instead of 'it'.

They don't like this for good reasons: it smacks of anthropomorphism. It's easy enough to ascribe personalities, and hence free will, to cars, trains and boats, even though we know that they are simple machines with stereotypical behaviours. Clearly there is a serious risk that we will read too much into the behaviour of non-human animals too. To do the same for a robot is just asking for trouble. On the other hand, quite apart from my annoyance at the arrogance that leads human beings to think that they are the only special beings in the universe, I'm trying my very best here to make a 'somebody', not a 'something' and this is why I refer to Lucy as 'she'. One has to have the courage of one's convictions.

So far, though, referring to Lucy as if she really had a mind of her own is pushing it, to say the least. What I've described in the previous chapters is a machine that is capable of organising itself into a structure that can point at bananas. She (it) can point at a banana, or she (it) can point at a banana; those are her only choices. Sometimes it's true that she doesn't point at the banana, but even at my most charitable I can only conclude from this that she makes mistakes, not that she sometimes chooses not to co-operate. Where are the signs of volition, of deliberation? Do we have enough yet to suggest that a genuine sense of agency will eventually arise if we simply add more cortical maps? Where might such coherence and self-determination come from, that we might one day feel justified in calling one of Lucy's descendants 'she'? It's time to bring this section to a close with some thoughts for future research.

The architecture that I've described in the previous chapters (not all of which has yet been emulated in Lucy) comprises a network of cortical maps, connected to sub-cortical structures such as the superior colliculus. Many of these sub-cortical modules are complete sensorimotor systems in their own right. The superior colliculus controls a reflex that orients the eyes towards a source of motion, while other structures are capable of executing small sensorimotor programs elsewhere in the body. In the absence of any cortical tissue at all, we still have the capacity to respond sensibly to many sensory stimuli, just like a frog.

What cortex seems to be doing is *extending* these reflex responses to a higher level. Except that this time, instead of being hard-wired, the responses are self-organised, learned and far more conditional. I've suggested the possibility that V1, for example, uses edge detection to extend the saccade-towards-movement reflex of the superior colliculus into a saccade-into-bounded-surfaces response. The Frontal Eye Fields and parts of the supplementary motor area of cortex might then extend this chain still further, adding even more sophisticated eye control. In fact it is possible to witness this hierarchy of responses in action for yourself. If you focus on an object with some purpose in mind, you tend to fixate strongly and saccade to specific features, ignoring any irrelevant stimuli, but if you allow your attention to wander slightly, you will find that your eyes spontaneously tend to saccade around inside the edges of the object's surfaces. Then if you deliberately defocus your eyes (hence preventing V1 from detecting any edges), you'll discover that you tend to stare blankly into space and saccade only if some sudden movement occurs in your peripheral vision. The same principle applies to other motor programs, such as reaching out and grasping something. It is possible for brain damage or surgery to destroy the top-down control signals that normally suppress or release this reflex, leading to an unstoppable urge to pick things up on sight.

Conscious will seems not so much to command action as to enhance, suppress or co-ordinate actions that otherwise would be spontaneous. Nevertheless, there *is* some control. Our minds are not centralised units at the top of a hierarchy of command, but neither are we just a collection of cobbled-together reflexes, reacting in a wholly bottom-up, uncoordinated and involuntary fashion to incoming sensory stimuli. I'm suggesting that these top-down influences flow along what I've been calling the yang pathway, and we describe them as attention, anticipation or intention, according to context.

When you concentrated on a visual object in the previous exercise, you suppressed, stimulated and guided your otherwise spontaneous eye movements by focusing your attention on to various parts of the scene. Just thinking about a portion of the image was enough to trigger an appropriate eye movement. When you see a car disappear behind a wall and your eyes move to where it will be when it emerges again, then you are most probably using the same kind of neural signals, although this time it would be more appropriate to call the effect prediction or anticipation, instead of attention. When you deliberately roll your eyes, as a form of non-verbal communication, the best word to use for the blob of nerve activity that commands this particular movement is 'intention'. These are three different kinds of eye movements, one attentional, one anticipatory and one intentional, but to me they all seem to represent the same neural process at work, albeit driven by different parts of the brain.

Incidentally, another term that might justifiably be included in the lexicon of yang signals is 'working memory'. When you put down an unfinished cup of coffee you keep a short-lived and fragile memory of where you put it, so that you can pick it up again without needing to look. If you move your body, your brain will automatically correct for the new relative position of the cup, suggesting that this memory is stored in or connected to a map that knows how to produce the necessary coordinate transforms. If someone distracts you, you may forget where the cup is and have to look around for it next time you need it, showing that working memory is either volatile and needs to be constantly refreshed, or is a limited resource that gets overwritten when something else takes control of the same map. When I asked you to think of windmills and tilt your head in chapter 16, it may be that your brain formed a working memory of several key locations on the image, in a map that has developed the ability to perform rotations that compensate for head movement. It has also been shown, by means of electrode recordings, that an intention to act (specifically, an intention to reach out in a certain direction) reveals itself as a build-up of nerve activity at the appropriate location in a cortical map, some time before the event itself. When the action finally takes place, the nerve activity disappears. The best interpretation of this experiment would require us to blur the distinction between working memory (a note of where the subject intended to reach) and the intention itself.

Where do all these top-down signals come from? They almost certainly don't emanate from a single, central, 'executive' module. It seems to me that they must arise at all levels of the hierarchy. Upcoming yin signals from the senses will eventually resonate with downgoing yang signals in such a way as to trigger a strong yang response, which eventually gives rise to movement (or some other kind of effect, such as a hormonal change). The more sophisticated the response, or the more sophisticated the sensory processing involved in recognising the right response, the further up the hierarchy the signals need to go before they are able to trigger a cascade in the other direction. This cascade may occur down the same chain of maps, as I hypothesised for V1's saccade-stimulating response to visual surfaces, or it may switch lanes and trigger changes in another pathway entirely, such as would happen when complex visual and social processing leads you to reach out and shake someone's hand. In this example, the sensory signals enter in V1 and run up the yin pathways some distance into the temporal lobes (to be recognised) and the parietal lobes (to be transformed into arm-centred coordinate space) before triggering a response in the frontal lobes that cascades down the frontal yang pathways, causing a sequence of arm movements to emerge from map M1.

The likelihood is that many such reflections are triggered simultaneously in different parts of the brain. Some of them can give rise to behavioural responses that do not interfere with the others but some will cause conflict. The responsibility for co-ordinating all these calls to action, deciding which are stronger or more imperative than others, and which ones need to be suppressed to avoid conflict between incompatible movements, seems to lie at least partly with the basal ganglia. It may be that the thalamus also plays a similar scheduling and releasing role, which I shall come back to in a moment.

It seems to me that the more sophisticated the processing involved in identifying or managing a response to a stimulus (or an expectation of a future stimulus), the more likely it is that the response will involve wholesale action on the part of the body. If we feel a fly land on our skin we can flick it off with a wave of our hand, even if we are riding a bicycle at the time – the response requires little sensory processing, little action sequencing and doesn't conflict with our other actions. Crucially, we may not even be aware that we have done it. On the other hand, a response to a predicted threat is likely to require enormously sophisticated processing to recognise it, judge its likelihood and pick the right moment to respond. The response itself is likely to be complex, and very probably requires the involvement of our whole body as a co-ordinated unit. So the further up the various chains of yin and yang we go, the more the action is likely to require us to behave as a coherent, unified entity, rather than an ensemble of independent machines. The more coherent and unified our actions, the more we feel them to be conscious and voluntary.

Leaving aside the appallingly difficult question of what consciousness really is, have we at last begun to see *where* it is, and why it operates at some times and not others?

Let's suppose, for the sake of argument, that the messy network of wiring between cortical maps is really a neat, inverted tree shape. Each branch and sub-branch of the tree conducts sensory yin signals, flowing from the leaves towards the root node of the tree, and attentional, anticipatory or intentional yang signals, which flow the other way. Each leaf or branching point of the tree is a cortical map, and the leaf maps connect to the various sensory and motor structures of the brain. This is a somewhat misleading simplification, but go with it for now. Imagine sensory stimuli rising up from various leaves, merging at higher nodes and then, finally, triggering one or more downstrokes of yang activity that give rise to movement. A nice metaphor is the way that several tentative feeder strokes rise up from the surface just before a lightning strike, triggering a much more powerful bolt that slams back down the shortest of these paths into the ground.

We can suppose that the yang signal – the lightning downstroke – is a limited resource. If a higher map sends a blob of nerve activity down the yang pathway to a lower map (signifying an attention shift, an anticipated future sensation, a working memory or an intended state that acts as a target for servo action), then this blob will take over the map entirely. No other blobs, recommending different actions or expectations, are permitted. For many maps, such an overlapping set of patterns would lead to conflict and a confused response, so if a map has been told to do (or expect or attend to) one thing, it cannot simultaneously be ordered to do another.

If a sensory stimulus rises only a little way up the hierarchy before triggering a response, then the rest of the brain is free to go about its own business. But what if the feeder stroke rises right to the top of the tree? Whatever response it triggers will now be in complete control of all the responses of the various lower branches. No other parallel or higher response can be initiated, because this first response owns the highest node in the tree and hence is the only signal that can propagate to any of the lower nodes. Would it be reasonable to say that this signal has thus been brought to our conscious attention? Have we become centrally conscious of the outside world, and 'decided' to make a unified and voluntary response? I find this idea very appealing.

On the other hand, if the sensory signals from the outside world do not need to rise very far up the tree before triggering a downstroke in response, then the higher echelons of the tree are clearly not being used for attending to our immediate needs. Our actions are being handled unconsciously or subconsciously. So what is going on in the higher regions of the tree when the lower regions have got everything safely under control?

One of the rules of thumb I've developed from working with Lucy is that at least one layer in each of her cortical maps must show a distinct tendency to exhibit spontaneous activity. I've also often found that it is useful to have neural layers that adjust themselves in such a way as to maintain a constant overall activity level. In other words, in the absence of any external signals, all the cells in the layer will fire to some extent, but if the activity level is forced to rise in one small region of the map, it automatically reduces everywhere else to compensate. This kind of structure has all sorts of nice properties, including the ability to generate predictive wave motion (imagine briefly pushing down on the water at one spot on a pond: everything else lifts up to compensate, since water is incompressible, and then tries to return to normal when you let go, leading to the creation of an expanding ripple).

If this phenomenon is present in real brains, then every cortical map *must be thinking about something all the time*. Sensory information or down-coming yang signals may *change* the pattern of activity on the map, but in the absence of influences to the contrary, the map will bunch up into a pattern of activity all of its own, influenced by its own neural geometry and recent patterns of activity (that is, its history). Working memory, of course, would be a natural consequence if such patterns had a tendency to retain their shape for some time after the incoming stimulus has been removed.

The existence of such self-organising but spontaneous activity would suggest that the higher portions of the tree are still busy thinking about *something*, even when all the body's responses to sensory stimuli are being managed lower down the hierarchy. And isn't this how consciousness feels to us? When we are awake we are always conscious of something. Sometimes we are conscious of the external world, but if nothing much is happening we spend our time in a day-dream, oblivious to our senses, often thinking about the future and playing 'what if' games to ourselves. Every now and then these internal machinations might build up to a level that triggers a large downstroke of yang, which reaches right down the chain and causes action, over-ruling any sense-generated responses. We have thus made an autonomous, 'central' decision, and we act upon it independently of what is occurring in the outside world.

The best place to see all of these levels of activity in action is when you are driving. Quite often you can drive for miles without being the slightest bit conscious of what is happening on the roads. You are in a world of your own, day-dreaming away, while the lower echelons of your brain deal with the mundane tasks of steering and predicting the intentions of other drivers. Only if something unexpected happens does the yin signal rise up far enough to command most, or all, of the cortical tree, demanding high-level cognition and a conscious, full-body response. Or, you may be musing happily away to yourself up at the top of the tree, when the thought strikes you that you are hungry and would like to stop and eat. The automatic programmed responses that otherwise would have carried you home without disturbing these higher thoughts are now subverted to the task of turning off the road and seeking out a restaurant. Your consciousness has reached down to the very toes of your brain and taken complete control.

This is all very speculative, even if supported by a long list of tantalising clues, but it does feel right to me, or at least, feels useful from a robotics point of view. It explains our perception of a central 'I', without requiring all decisions to emanate from a master controller. It explains why we are sometimes conscious of the outside world and sometimes only of our own imaginings. It explains why we can do some things simultaneously while others command our whole being. The problem with the idea is that it implies a tree structure, with a root node, whereas my initial hunch for what lies at the 'top' of the much messier and more intertwined cortical network was the hippocampal system, which can hardly lay claim to being a master controller or the seat of our highest mental existence. I'll come back to this in a moment, since I have a few more things to say that may help to make sense of this.

Let's start with the hippocampus. Towards the edges of the more sensory lobes of the cortical sheet, where it wraps under itself like a curling leaf, the complex six-layered structure of cortex connects with an evolutionarily older, three-layered structure. This is subdivided into various regions, but the whole thing is known as the hippocampal system (HS). The HS takes input from and returns output to many higher maps in cortex, forming a signal loop that sounds just right for implementing a final mirror, which reflects any yin signals that have made it that far back down the yang pathway.

One of the things the HS seems to be particularly good at is making very rapid connections between incoming signals, which seems to underlie its role in the formation of episodic memories. The first time that we experience something, we tend to remember it as a specific event: an episodic memory. If you had just visited the dentist for the first time, you would remember the experience as a specific episode, associated with a particular moment in time and a particular context. On the other hand, after you have been to the dentist a few times, the experiences tend to blur into each other and become semantic memories: knowledge of what a visit to the dentist is like and what it means to you, without the memory being localised in time and space. It seems that the HS is responsible for the first of these types of memory, and given its hypothesised position at the top of the stacks of cortical maps, it isn't too hard to imagine how such episodic memories could be formed by aggregating together other memories, stored lower down the cortical chains.

We don't need to remember the precise sequence of raw sensory stimuli in the way that a video camera does – to do so would require absurd amounts of storage. Instead, we can temporarily tie together smaller, invoked memories of what a white coat looks like, what a whining noise sounds like and what sheer terror feels like. To reconstruct the events we can then literally re-member these fragments of memory into a complete story. (Incidentally, from this it is easy to see how 'false memory syndrome' can exist; how people can have memories planted in them by suggestion in such a way that they link up chains of events that individually are familiar but never occurred together in reality.)

If an episodic memory is a conjunction of existing, rather more semantic memories, then where do these semantic memories come from? There is strong evidence to suggest that memories 'migrate' from the HS into the neocortex over a period of days. It sounds to me rather as if the hippocampus forms highly associative, specific and rapid memories, which later go on to influence the structure of the rest of cortex, producing a more statistical change in one or more cortical maps before the original memory finally fades away. Such statistical knowledge then provides the construction kit from which new episodic memories can be assembled in the future.

If this is the case, we can assume that the memories tend to migrate to the lowest level in the cortical tree that can support them. This movement of an episodic memory down into the correct parts of neocortex might even take place through pure trial and error. Imagine there are horizontal fibres in the cortex that constantly try out new links and associations. If a new situation arises that bears a lot of similarity to one that has previously only been experienced once (and hence lies stored in the connections of the hippocampus) then the yin signals that carry the sensation of it must rise all the way to the top before they are able to trigger a useful response. However, if some of these neocortical association fibres manage by accident to duplicate part of the essence of this memory, the new sensory signals might no longer need to travel so far up the tree before being recognised. Over time, the original memory will fade from the higher areas (which never get invoked because the response is now triggered from lower down the hierarchy) and the trace will migrate lower and lower until it finds itself at the lowest level capable of representing the essential idea. Such a mechanism might help to explain why we need to be fully conscious of what we are doing the first few times that we do it, but can respond completely unconsciously once the essential elements of the experience have been learned.

On the other hand, the hippocampus doesn't sound like the top of a chain of command, just the top of the search through memory for a useful recognition or response to events – a final, fully reflective mirror above the stack of cortical half-silvered mirrors. It sounds like there may be another kind of 'top' to the system as well, representing the most abstract, long-term and deliberate levels of thought.

When a cortical map that is close to the primary sensory input or motor output makes a prediction, it is probably thinking a mere few milliseconds ahead of reality, anticipating the next step in events. Somewhere deeper in the tree, exactly the same mechanism must be making predictions that extend a considerable time into the future. When we respond to social situations we may make decisions whose effects will not play out for many years ('will you marry me?'). Such long-term intentions and predictions count as the 'top' of a hierarchy, in the sense that they give rise to rather more immediate predictions or motor commands, which in turn lead to low-level shifts in attention or primitive muscle movements. They also count as the 'top' in consciousness terms. Not only are we generally fully aware of these broader contextual aspects of our sensory lives, while we often haven't a clue what body movements we need to invoke in order to act on them, but also we are capable of reflection at these abstract and long-term levels in the absence of sensory signals. When we do this we are consciously imagining things, and when we put together chains of possible actions and their anticipated consequences, without immediately carrying them out, we say that we are making plans.

The most likely location for these higher, volitional maps of association is the front of the frontal lobe; the other end of which contains M1, the primary motor cortex. Nevertheless, I think it makes some sense to visualise the various lobes of the brain as loosely gathered sticks in a camp-fire, in which case each stick has its own top level and there is no single point where everything comes together. To some extent, therefore, I think there are various tops to the system and it is even vaguely conceivable that the relative dominance of these different sticks helps to define our personality and mode of thought.

Musicians, mathematicians, linguists, painters and engineers seem to think very differently from each other. I sometimes feel it's a miracle that we can communicate at all, because internally we seem to process the world in radically different ways. My wife thinks in words, or at least, words are how her thoughts manifest themselves, and she finds it very hard to think in pictures. My son and I think mostly in mechanical terms: the laws of physics automatically animate the objects we see with our mental eyes. I can look at a computer program or a machine and quickly grasp its properties but I can stare at an equation until my eyes explode, without it speaking to me at all. On the other hand a mathematician will see the full implications of an equation in a single glance, but might not be so good at thinking in a procedural way. It occurs to me that there might be an anatomical basis for this: if you pick a spot on any cortical lobe that is as far away as you can get from the primary inputs and outputs, you are likely to happen across the highest centres for handling one aspect of thought: body image, language, mathematics or vision, say. It would be silly to speculate too far, but it does seem possible that we each learn to favour one particular stick on our cortical camp-fire over the others. Perhaps we could usefully characterise someone like me as a 'rostral parietal lobe thinker', or a mathematician as being 'medial temporal lobe dominant'?

What about this idea of abstract symbolic thought, the sort that mathematicians often favour and that frequently makes it difficult for me to communicate successfully with physicists? The earliest and still perhaps the strongest motivation behind AI was to automate the symbolic operations that seemed to underlie the best of human mental processing (or at least seemed that way to mathematicians). Researchers often regarded sub-symbolic mental activity as having little if anything to do with intelligence, and of no significance to their endeavours. More recently there has been a backlash that pooh-poohs the whole notion of symbolic manipulation and denies the existence of any form of abstract representation of the world inside the brain at all. Clearly we do a lot of our highest thinking without recourse to abstract symbols, but equally clearly we are capable of abstract reasoning when we need to be, so the truth must surely lie somewhere between these extremes.

Since I'm speculating wildly and will by now have caused anyone who would be offended by this thought to slam the book shut and miss what I'm about to say, I will offer the suggestion that perhaps the brain really does manipulate symbols, some of the time, and that these symbols act rather like internal senses.

We have seen that spatial coding – the creation of patterns and blobs of nerve activity on the surface of maps – is probably a universal feature of cortex, if not of the brain in general. Some of these maps use very concrete coordinate frames, for example the retinotopic arrangement of the early visual maps or the homuncular (body-shaped) arrangement of the somatosensory and motor maps. Some are slightly more abstract, as in the tonotopic maps of the auditory cortex. In some cases at least, blobs of nerve activity may essentially form counters or tokens that can be moved around these maps, representing the focus of attention, the location of something in working memory or the goal state for a servo action. Occasionally, we might even say that the computations are conducted by the blobs and patterns themselves, rather than by the neurons. In this sense, it is fair to say that the data the brain computes with are not tied to the neurons that transmit and modify them. They are coherent and meaningful entities in their own right, and are mobile, being able to move around the cortical maps even though the neurons themselves do not move.

Suppose some of these maps act like working memory, by which I mean they can maintain their activity patterns in the absence of input. And suppose that some of them have learned (not evolved, although evolution could easily make certain kinds of learning more likely than others) to represent highly abstract and symbolic coordinate frames. Imagine how mind-boggling it must have been to be the first person who figured out how, by moving pebbles on a beach and drawing lines around them in the sand, one could add up, share out and perform the other basic operations of mathematics. Such operations require movable pebbles or other tokens, plus a meaningful space in which to arrange them and a set of transformations that converts one pattern of tokens into another. Cortical maps, at least in my interpretation, do precisely these things. Just suppose we evolved enough cortical real estate to spare a few maps that weren't connected to the outside world. In fact imagine them to be like an internal world, a patch of pebble-strewn sand. Creating dead-end streets in the cortical wiring that develop their own emergent coordinate frames doesn't seem impossible for a tension-and-balance morphing process, such as the one I outlined in chapter 17. By treating such arenas as if they were primary sensory/motor maps, we could both perceive and manipulate this internal world of symbols in just the same way that we handle the external world of sensations and movements. Perhaps this is how the brain developed the ability to make abstract symbolic manipulations, such as those used in arithmetic and language?

One final piece of idle speculation before I get too carried away: None of the things I've said so far are quite enough to explain *thinking* – the stepwise, coherent manipulation of ideas and their transformation through various processes. As I've mentioned before, if I ask you to work out whether the letter 'p' is the same shape as the letter 'd', you need to rotate one image and compare it to the other. It is inconceivable that your brain was able to grow and maintain a cortical machine specifically for answering such questions, and pure symbolic manipulation doesn't begin to describe the task, so I presume you must be making deliberate use of one or more cortical maps that have wired themselves up to perform rotations and comparisons for other, quite unconscious, reasons. To do this, it seems to me, you need to be consciously able to switch information around from map to map, creating new production lines on demand for manipulating internally or externally generated information.

The thalamus is a structure that I've only mentioned in passing, but most cortical signals pass through the thalamus on their way from map to map, or on their way between cortex and the senses or motor systems. It is likely that the thalamus modulates these signals and forms an integral part of a thalamo-cortical system of resonating (yin/yang) loops. Indeed some scientists regard the thalamus as the seventh layer of cortex, and given the vast numbers of inter-cortical fibres that take up room in the cerebral hemispheres, the cortex does look remarkably like an outer shell of the thalamic bulb, which evolution has had to peel off and blow up like a balloon, purely in order to allow room for the intervening white matter.

The thalamus is also a set of maps. The signals from the senses arrive in logical (retinotopic or homuncular) order and array themselves on to the surface of the thalamus. The various cortical maps also make one-to-one mappings to the thalamus, so all the assorted coordinate frames of the cortex are mirrored on the thalamus. Just suppose a part of the cortex, or the thalamus, or some other structure, somehow developed the ability to wire itself up into a map of all these maps. In other words, suppose one map in the cortex was set out as a representation of how the other maps (including itself) were arrayed across the thalamus. It seems to me at least vaguely possible that this innovation would allow one cortical map to control the flow of data into and out of all the other cortical maps. It reminds me of the diagrams of the railway network you find in modern signal boxes, upon which are mounted buttons that allow you to control the flow of trains around the real railway. If such a thing were possible, it might allow one cortical map to learn to control the routing of signals around the other cortical maps. Indeed, this cortical 'map of maps' could control itself, in a highly self-referential and self-modifying manner.

A general-purpose digital computer is nothing more than a set of switches that can turn each other on and off. Combine this idea with the previous thought about symbol manipulation arenas and you may see the first faint hints of a very interesting possibility – a mechanism that might go some way towards explaining the conscious, abstract mental processing that is characteristic of human beings (and possibly a few other species). Could such a mechanism arise from such a simple evolutionary modification to the mammalian cortex? As always, I haven't the faintest idea, but it is worth thinking about.

Do you think these ideas make some kind of sense? I do, more or less. Unfortunately, in order to explain them I have needed to wave my hands in the air a lot and use loose language. Computers don't understand loose language – you need to explain everything to them in words of one syllable, because they are very, very stupid and very, very literal. Turning all these 'sort ofs' into precise wiring diagrams and rules is going to take a great deal of work, but I hope to be able to test out some of these more nebulous and far-reaching ideas inside Lucy's brain over the coming months and years. Whether any of it actually holds water remains to be seen.

My main difficulty, unsurprisingly, is that I am still confused by so much. The ideas I have outlined in this book seem to explain a number of things that initially puzzled me, but the brain remains a dreadfully counter-intuitive source of evidence. To take an example completely at random: why is it that you can ask me a difficult question ('in which year did Lock, Stock and Barrel publish their famous paper on such-and-such?') and I will know *instantly* whether I know the answer or not (and hence whether it is worth thinking about), even though I might not actually be able to come up with the answer until much later? How can I know that I know, without first having recalled the answer?

Perhaps I had better ask an academic . . .

PART FOUR
Spirit

When people ask me questions about this project, they usually aren't really interested in artificial intelligence, *per se*. To some extent, neither am I. What they are really interested in is themselves and what AI may or may not have to say about the human condition. They worry about how AI might alter their future, or their children's future, and they worry about what the idea of machine intelligence says about the things that they hold dearest, such as consciousness and emotions. And I don't blame them.

I don't think we can explore these issues without first understanding quite a lot about natural as well as artificial intelligence. I detest issue-driven popular science – understanding is a prerequisite for making responsible decisions and we have to get everything in the right order or our understanding will be coloured by the issues, rather than the other way around. Moreover, I think the way we view these things depends enormously on what kind of AI we are talking about, what we believe is possible and how we believe it can be implemented.

The bulk of this book was written with these factors in mind, but finally I'd like to turn to a few of the implications of this sort of work.

CHAPTER NINETEEN

Misbehaving

Yesterday (and I'm writing this chapter wildly out of sequence) I did a short interview on the radio. It was one of those news magazine pieces, designed to take a complex and difficult issue and scrunch it down into a sound bite for commuters to mull over while they are waiting for the lights to turn green. It allows us all to feel informed and concerned without actually having to deal with the messy complexity of other people's problems. I wouldn't have bothered, but it was a subject that gets my blood pressure up, so I reluctantly agreed to take part.

'Will intelligent machines take over the world?'

'No.'

'Should we urgently set up an ethics committee to monitor developments in AI for public safety?'

'No.'

'But ...' (and here my antagonist pulled out his trump card; the one subject I'd dreaded having to deal with in such a short time) '... what about military applications? Suppose someone develops intelligent missiles with minds of their own that we can't control?'

Time was running out and I had to make a quick response. 'But we already have intelligent missiles with minds of their own,' I protested. 'They're called pilots.'

This piece went out in the early morning news programme. Travelling up to a studio and back in the rush hour used up most of the morning and I'd lost the will to think by the time I got home, so we turned around again and went shopping. About mid-afternoon, we were browsing in the menswear section of a department store, when we noticed rather a lot of activity in the area where they sell television sets. Instead of the usual handful of men waiting for their wives to return from the toilet, there were dozens of people, simply standing, staring into the middle distance. It was unnerving. We found ourselves moving towards them, drawn by the intensity of their introspection. Not a coherent crowd, just clusters of lonely people sagging silently amongst forgotten carrier bags. And then, on the ranks of television sets, repeated over and over as if one picture wasn't enough to do it justice, we saw the unmistakable skyline of downtown Manhattan. But it was all wrong: something was in the picture that shouldn't have been, and worse still, something that should have been there, wasn't.

Yesterday was September the 11th, 2001.

I couldn't make up my mind whether to begin this chapter in this way. I hope it's not disrespectful. I have nothing of value to say about what killed those thousands of people, nor what we should do about it now, nor what we should have done about it long ago. Not in this context anyway. But this book is essentially a diary, and to pass over yesterday without comment would be to deny that it happened. Even if I were writing a cookery book I'd feel I had to say something. And I can't help but recall the repulsive irony of these events in relation to my offhand comment about intelligent human weapons, just a few short hours beforehand.

I'm sick of people asking me when intelligent machines will take over the world. I'm sick of being thought of as a megalomaniac mad scientist, perpetually bathed in a green glow as I attempt to conjure up a race of golem warriors to avenge me for being bullied as a child. I'm sick of people emailing me to tell me I should think about what I'm doing, as if I'm just fooling around on the strength of an idle whim. I'm not feeling as sick about these things as most people on the planet are feeling about something else this morning, I grant you. Nothing that really mattered the day before yesterday seems to matter much today. But life must go on, where it can, and I really need to tackle this subject while it is fresh in my mind.

To be honest, for once I really wish I *could* conjure up a hyper-intelligent master race. Perhaps they wouldn't show much sympathy for morons like you and me, but they would surely make better stewards over their dominion than we do. I don't fear intelligence. On the whole, intelligent beings are thoughtful and insightful.[1] They actively seek understanding and wisdom and they are acutely aware of themselves and their effects on others. No, the thing that makes me afraid, above all else, is stupidity. The real threat to us and to all that depend on us, is people who don't use the brains they were given, and prefer to follow the herd or act short-sightedly.

Especially, to be brutally blunt, those people who justify their actions in terms of someone else's ideology, because they simply can't be bothered to think for themselves. Yesterday's terror seems to have arisen at least in part from an unquestioning adherence to, or perhaps more likely (although this is just as bad) a post-justification by, a book of rules called the Koran. Today I've been shuddering at the number of times I've heard American voices (and less often British ones) appealing self-righteously, almost threateningly, and apparently without a moment's serious consideration, to their own brand of religious rule-book, the Bible. As if it were possible to codify the rules for living. As if a moral code – one size fits all – is a substitute for moral *principles* and personal responsibility. It seems the only decision some people make in their entire lives concerns which of the available creeds to adopt, and even on this question they prefer to take advice.

If people want to believe in divine beings, that's entirely their business and I also don't deny that most religions preach tolerance and moral rectitude. But it's also in their very nature that religions tell us what to do. They make us sign a blanket acceptance that we will adhere to the creed, rather than think things through for ourselves. It is perfectly possible to be a moral and upright being without joining an organisation and donning an intellectual uniform, but that's not how religions work. They cause us to switch off our intelligence and behave like armies of centrally controlled drones. Exactly the concept that, when applied without much thought to robots, frightens the living daylights out of people.

If it isn't a religious ideology that people rest upon, it's usually a political one, or the newspapers. We absolutely love to have someone

[1] Perhaps you know some intelligent people who are haughty and arrogant, or even malicious, but that's not the point. Take this test: you are in a narrow dark alley. To your left is an alligator. It's not much bigger than you and not inherently malicious, but has very sharp teeth, no moral scruples and just enough brainpower to know that you are edible. To your right is an even more potent monster – a ton of solid metal, capable of exceptional speed. It's a car, driven by a highly intelligent professor of English literature. Oh no! Which way do you turn?

tell us what to think, and we read the paper that tells us what we want to be told. It seems to me a remarkable fact that here, in the West, we routinely fell millions of trees, devote huge energy resources to boiling them down into newsprint and then cover them in toxic metals and organic salts, so that thereby informed people can drone on in the pub about how we're ruining our environment by cutting down trees, wasting energy and all the rest. In a sane world, you'd imagine this would trouble responsible people, but of course no newspaper is going to tell us to think in this way, so we don't.

None of this applies to you, of course. I'm fully aware that I'm preaching to the converted. Nor is it my intention to insult journalists, who I've already explained are intelligent, thoughtful people. I'm simply pointing out how insidious memes, such as religions and newspaper-enhanced ideologies, can promote inflexible, unthinking behaviour even in human beings, and how much more dangerous stupidity is than intelligence.

I hope to goodness this book isn't informative. I mean, I hope that it raises more questions in people's minds than it answers. All I have to offer are my own hard-won opinions, but I don't want to present them as a *fait accompli*. I would rather that people struggle alongside me as we battle with the problems than expect me to spell out the answers. Luckily, Lucy is certainly a journey more than she is a destination. I specifically hope I can disturb people into wondering what it actually means to be alive and to have a mind and to realise that this is something we can think about for ourselves, not just leave to the experts. Artificial life is valuable because it throws our cosy certainties into disarray. We see our own reflection in a new light. Our neat little categories – us versus them, man versus machine, living versus non-living, conscious being with moral rights versus mindless walking beefsteak – all crumble before our eyes and we are left with a healthy uncertainty; a lack of easy answers that obliges us to think things out afresh, without the help of dogma. All these questions have been with us ever since we developed the capacity to ask them, but we didn't ask, or if we did, we didn't have any other examples to work with that might turn the spotlight of our examination into a broader beam. Actually, now that I come to think about it, perhaps it's not the fear of violent machines that really troubles us at all; maybe this is just a smoke-screen to hide the fact that we are scared to face up to the question of what we are ourselves.

Oh well, I can't create a master race anyway. In fact the whole concept of hyper-intelligent megalomaniac machines running amok arises largely from a series of innocent mistakes and misapprehensions. I talked about these to some extent in my previous book (*Creation: life and how to make it*), but you may not have read it and anyway, since then I've had the chance to listen to more people's views and think more about my own, so I shall tackle the subject again here.

When you think about it, the main reason the issue arose in the first place is simply that it offered a good story-line for science fiction writers. *Invasion of the Terribly Thoughtful, Intellectual Robots* isn't exactly a snappy title and it doesn't leave much room for a plot. Of *course* it all goes horribly wrong in novels – how could the hero triumph in the end if there is nothing to triumph over?

Strangely, we don't worry about any other kind of intelligence taking over the world. Suppose a naturalist were to return tomorrow from the jungles of Borneo and proclaim the existence of a new species of primate. Would our first thought be 'oh my goodness, I wonder if it is going to take over the world?' Of course not. We only think this about robots, and then not about real robots, only the hypothetical sci-fi robots of our self-titillating nightmares. We already share this planet with tens of millions of species of creature, and let me tell you: every single one of them is determinedly trying to take over the world. Every daisy and every butterfly on the planet is furiously and frantically trying to spread its seed as far as it can and is doing its utmost to swamp the competition. But do we worry about it? No, for the most part it remains in balance. Adding a few artificial species to the millions of natural ones isn't even enough to compensate for all those species that we thoughtless humans have wiped out.

I can't deny that human beings are currently doing rather better than the daisies, when it comes to taking over the planet. For a comparatively short but perhaps critical moment in evolutionary history, a species has developed far more power than it understands what to do with, and the balance of nature has been shifted sideways by the impact. We must cross our fingers and hope it doesn't topple. Intelligence undoubtedly lies at the heart of this potential calamity, because it was having the intelligence to construct tools, and the concomitant development of complex cultures, that put us in this unstable position. But to blame it on intelligence *per se* would be an error. As a species, our best way out of this fix is almost certainly to improve our intelligence, not diminish it. We need to understand more, to know more, to develop better skills, greater foresight, better empathy. All of this requires us to continue honing the gifts evolution has given us. If we are smart enough, we can correct our initial inexpert wobble.

Very little of this applies to artificial intelligence at all. Despite some pundits' wishful thinking and other people's bizarrely masochistic hopes, we are *nowhere near* to creating anything with the intelligence of a human being and even if we were, there is no reason to suppose that these creatures would either grow smarter than us quicker than us, or decide that they envied the human race enough to want to attack us.

The first big mistake, which seemed far more plausible in the nineteen-sixties than it does now, and explains why most of the big AI novels and movies arose when they did, is to assume that intelligence is something that can be abstracted out and streamlined. This comes from the school of thought that tries to emulate the mind rather than the brain, as I outlined in chapter 2. People assumed that there was a close correspondence between what minds do and what computers do, and that relatively simple sets of mathematical processes could emulate the reasoning power of the human mind – sufficiently so, in fact, that we would be able to program a computer with these rules 'any day now'. By extrapolation, of course, and thanks to Moore's Law, if the computers of 'any day now' were fast enough to create human level intelligence, then the computers of the day after tomorrow would presumably be capable of supporting superhuman intelligence.

This theory falls down in every respect. We have so far failed to find any means by which mental processes can be generally represented in terms of streamlined mathematical or logical rules, and 'any day now' keeps passing without significant further progress. We neglected to recognise that logical reasoning is the easy part of what minds do: that moving around a room, recognising an object by sight or reaching out and picking something up are horrendously difficult problems by comparison. We failed to capture the essential fact that intelligence is not just reasoning – not by a long, long way.

It isn't simply a question of tackling each aspect of intelligence one step at a time either, as some suggest. There is no evidence to support the idea that intelligence, and especially not consciousness, can be put together in such a piecemeal way. Perhaps we can figure out how to achieve visual pattern recognition in a computer-friendly way (although we certainly haven't yet). Perhaps we can deal with navigation through space in a different way and bolt on a pattern-recogniser to enable a robot to see where it is going. Perhaps we can deal with motivation separately from knowledge acquisition and in turn, semantic memory can be dealt with in isolation from procedural learning. But now it is getting much harder to see how these might be bolted together into any unified and coherent entity and even if we managed it, what would the robot (I don't feel able to call it a creature) actually want to do? Whenever people try to divide up intelligence in this way their flowchart always seems to have a few boxes optimistically captioned 'to be looked into shortly' – minor things like consciousness, empathy and creativity.

These psychologically divisible concepts are not separate systems in real creatures, but different aspects of one system: the brain. I may be wrong, but I think I have good reasons to suppose that this sort of kludging together of separate sub-intelligent systems will not and cannot work. If you don't trust my opinion on this (and that's entirely your privilege) then I challenge you to make it work. Nobody else has managed, no matter how imminent they might claim success to be when filling in grant applications. It seems to me that a serial digital computer is simply an inappropriate vehicle for making a mind. I hope I've already shown that computers can be used as a valuable source of virtual materials for testing out ideas, and it is just about feasible to make the rudiments of a mind in this fashion, but minds are inevitably and invariably and crucially a consequence of the operation of a massively parallel system, and even if Moore's Law holds good for many decades to come (and there are reasons to suppose it will not so much break down as become irrelevant in the medium term) a serial digital computer is nowhere near big enough to emulate the behaviour of a hundred thousand million neurons and their one thousand million million synapses.

Moore's Law is a red herring anyway. If we are to make usefully intelligent artificial life forms, with brains containing even a few million neurons, then we need an entirely different sort of substrate from the digital computer. It seems very plausible that we will develop (once we know how to wire them up) silicon chips or other direct implementations of massively parallel neural networks. Perhaps we will develop the ability to grow and use real biological neurons first, but even though that feels like a whole different ball game in several respects, what I'm about to say still applies. Suppose we were able to build an artificial brain the same size and essentially the same design as a human one. If we then built one twice as big, would it be twice as intelligent?

Probably not. When you look at the whole of evolutionary history, you do find a correlation between brain size and intelligence. It stands to reason that having no brain at all is going to limit your options, so some kind of proportionality is only to be expected. To be fair, we should do our sums on the basis of brain size divided by body size, rather than brain size alone. Much as it pains us to know this, a sperm whale has a far larger brain than a human being, but mercifully it is smaller in terms of its body size. This is not an insignificant observation. Brains get bigger according to how many inputs and outputs they need. A sperm whale's skin is so vast that it needs a lot of neural real estate to deal with the huge number of touch, pressure and temperature sensors on its surface. Perhaps if you were to account for the unusually high density of sensors in a human being, such as the massive concentration in our tool-using fingertips, our language-generating lips and tongue and our high-acuity retinas (not to mention those in our genitals, which get more than their fair share of brain space), we don't have such remarkably big brains as we think.

In any case, at the level of individuals, brain size and intelligence don't seem to correlate well at all. Women have smaller brains than men on average but are no less smart (OK, they are smarter); people with hydrocephalus sometimes have almost no cortex at all but their diagnosis isn't usually triggered through any glaring stupidity. People with severe mental handicaps don't usually have smaller brains either. Having a bigger brain doesn't necessarily mean a greater level of intelligence. Moore's Law does not apply to brain tissue.

Appealing to evolution or faster chips doesn't work either. Evolution acts at a rate determined by the number of individuals, the length of their reproductive and life spans and the rate of change in their environment. These factors apply to highly complex machines just as firmly as they do to us, and so any fear that machines will romp ahead by evolving more quickly than us is borne of the same fallacy about intelligence being capable of abstraction and streamlining. The two hundred million years it took for the most primitive mammals to evolve into creatures like us is not a consequence of natural evolution being lazy or inefficient. Why should evolving machines be any different?

A similar but subtler argument applies to chip speed, too. Suppose the signals in a silicon neural network can travel thousands of times faster than those in our own neurons; won't that make them smarter than us? No. Maybe they'll always beat us at tennis, but they wouldn't learn any more quickly than us in general. It takes us a lot of trials to learn some things, but not necessarily because we are too stupid to 'get it first time'. We need lots of trials and practice to give us good statistical data. Anyone who learns things from a single example is in serious danger of over-generalising. When we see this behaviour in humans ('I cracked a mirror this morning and now my soufflé won't rise') we call it superstition.

In movies we sometimes see superheroes scan a book at ultra-high speed and digest its contents in an instant, and although the physics of our environment limits the speed at which we can perform physical actions (Hollywood is obviously exempt from the laws of physics), I imagine it might be possible for an artificial life form (or a future human) to absorb and digest pre-collected information more quickly than us. Even this is a bit of a long shot, though, because absorbing information is not the same thing as intelligence. We don't worry about the fact that our present-day database programs can learn more quickly and extensively than we can. Maybe faster signal speeds or more rapid access to knowledge would lead indirectly to higher intelligence, but not 'any day now', and it doesn't follow at all that this would result in the enslavement of humanity.

About the best argument I've heard in favour of such hypothetical artificial beings having an advantage over us is that they might live indefinitely and therefore have a longer education. Quite apart from the possibility that we'll achieve immortality for our own bodies more quickly than we solve the problems of AI, experience in itself doesn't create intelligence; it imparts wisdom. What, I wonder, do we have to fear from an exceptionally wise old robot?

One of the things that the abstract-and-streamline lobby presume can be left out is emotion. Future artificial intelligence, they implicitly believe, will be so much smarter than us because it won't be hampered by all those messy feelings. This is utter gibberish. Pure reason without emotion is what gave us Star Trek's Mr Spock. Pure emotion without reason, of course, is the province of Captain Kirk, and who is it that always triumphs in the end? Underneath this dichotomy lies an age-old obsession with the supposed tension between head and heart. 'Never let your heart rule your head' is probably good advice under many circumstances, but not all of them by any means. Taking any decision that involves other people, for example, demands that you consider how that decision might make them feel. You can and probably should do this dispassionately but you can only answer the question if you know how it would make *you* feel. This is where Spock usually screws up, because people (and aliens) don't usually behave in the way he expects them to.

Emotions are far more fundamental to intelligence than that, though. Why do we act intelligently? Because it enables us to survive. But how do we know whether something enables us to survive, without first having to discover that its absence kills us? How can we judge whether an action will be good or bad? The answer is part of the reason we have emotions. Emotions stand in for death, survival and reproduction, the things that rule life. We don't have sex because it promulgates our genes, but because it feels good. We don't pair up through a calculated risk assessment, but through love. We don't rashly step off a cliff to see what happens, because once we fell off a chair and became fearful of heights. We don't behave in socially acceptable ways to avoid breakdown of the tribe, but to avoid embarrassment. We don't eat to refuel our bodies, but because we are hungry. If we didn't have emotions then we would have nothing that we wanted to do. Emotions anchor us in the world. We may feel slaves to them, and we are, but we need them to survive, and so does an intelligent robot. You can have all the intellectual power in the universe at your disposal, but if you don't feel anything you will never be motivated to use it.

Which brings up the question of envy. It seems to be taken for granted that androids would envy humans and therefore want to take over. But does this make any sense? Humans certainly show envy. We always want to be somebody else – richer, warmer, more famous, younger, more popular, bigger-breasted, more interesting, Australian ... and why is this? It's because we aren't happy with our lot and to a very large extent that's because we haven't got used to it yet. We evolved in a very different world from the concrete jungle many of us inhabit today, and consequently our emotional structure is ill fitted for the way we live. Go to a place where you feel really calm and happy, or choose a landscape painting that makes you feel this way, and the chances are you will have picked somewhere that a chimpanzee would feel happy too. So much of our stress comes from the fact that our emotions are not well aligned with the needs of our modern lives, and hence we always want to be someone else who seems more comfortable (even though they don't see it that way and probably envy us).

Yet we rarely, if ever, envy gazelles or tigers or polar bears, so why are we arrogant enough to think that they would envy us? Give a polar bear a sub-machine-gun and the brains to use it and would he stride forward demanding a pinstripe suit and a Porsche, or insist on joining an exclusive leisure club? Would he heck! And this applies even more strongly to artificial life forms.

Suppose we wanted to make an intelligent robot to clean sewers – not a dumb robot that just blunders around but one that's capable of collecting useful items for recycling and knows how to manage the water flow just right. To enable it to learn how to perform its task, and for it to know how successful it is being, we need to provide it with an emotional substrate. Would we give it emotions like ours: a love of fresh air and a fear of the dark? Of course not. In the same way that our emotions are fixed in us by evolution (no matter how unhappy that may make us for much of the time) we would also have to fix emotions into the robot. We would want these emotions to be appropriate to the task in hand. We'd create joy whenever it did what we wanted it to do, and unpleasant sensations when it did something wrong or predicted that inaction would be undesirable (such as fear if the sewage backed up enough to threaten to overflow).

Toying with a creature's emotions like this might seem cruel and cavalier, but that's only because we insist on measuring such things on the basis of our own emotional responses. These creatures would not be repelled by having to live in a sewer; they'd be happy. What more could anyone ask from life? It seems a great deal more humane to me than sending boys up chimneys or yoking mules to carts. We are victims of our evolved emotions, but that doesn't mean we should inflict such woes on our creations as well. If a sewer robot is deliriously happy shifting excrement and gets a real kick out of fixing another blockage, why on earth would it envy us? Playing football or going clubbing would be anathema to it. We mustn't judge future artificial creatures by our own rules. For them to want to enslave us or even emulate us would require us to make the pointless and absurd mistake of building them to want that very thing. The only technological reason to build them at all is so that they learn to do (and hence must enjoy) some of the things we hate. Everyone's a winner.

Finally, I must return to weaponry. The one place where someone has a motivation to create artificial intelligence with malicious intent is for military purposes. This would, of course, be an appalling thing to do, if taken to extremes. Yet military experts are not despots, on the whole, so their motivation for making intelligent weapons would presumably be to prevent them from making stupid mistakes and killing the wrong people. Such machines, even if very intelligent (and high intelligence isn't always that desirable, hence the term 'cannon fodder') would still fit into the non-envy category in terms of not having any reason to 'break their programming' and do something we didn't intend them to do. Presumably they would consider it their life's ambition to perform their duties to perfection and blow themselves up with pride, but this is getting horribly familiar, here on September 12th, 2001. The real issue lies with who specifies the mission, not who or what carries it out. This is the crux of military ethics. Presumably a really hyper-intelligent nuclear missile would have both the brilliance and the integrity to turn itself on its owners, if it thought their actions were unjustified. In which case that would be their lookout. But there's no question that substantially dumber robot weapons have both good and bad roles to play, in just the same way that dogs can sniff out mines, dolphins have been used in anti-submarine warfare, and terrorists have turned airliners into missiles.

And therein lies the point: the evil of terrorism is not a reason to stop building airliners (or, for that matter, skyscrapers). Equally, the possibility of misuse by malign human beings is not a justification to stop work on artificial intelligence. AI is not like nuclear weapons research – bombs can only be used to blow things up, and so their manufacture is intimately tied up with their use, as are the ethical questions. There is barely an object in the world that doesn't have a military application. Even spoons and forks permit armies to march on their stomachs. Artificial intelligence has far more peaceable applications than it has military ones, and even the military ones are for military ethicists to think about, since the technology itself is neutral, and ethics only apply to applications. Artificial creatures are no more malicious than natural ones. It's human beings that we, and for that matter they, need to fear.

CHAPTER TWENTY

Descartes and his amazing ectoplasmic jelly

Yes, there is a god. Yes, when you die you go to heaven and sit on a cloud. No, robots can never be truly conscious and can never feel emotions like us. Yes, humans are the only species blessed with a mind. No, the tooth fairy will never forget to visit you.

Happy now?

I've lost count of the number of arguments, heated discussions and lectures I've been treated to about consciousness. Let me make it clear from the outset: I have no idea what consciousness is, or what it means to be aware. But I do have more than a sneaking feeling that my antagonists are looking in entirely the wrong place.

I've heard all sorts of deep and sophisticated arguments and almost all of them boil down to an implicit or explicit belief that consciousness is some kind of 'stuff' – something fundamental and indivisible. More often than not, it is portrayed as the true substrate of the universe. The idea is that consciousness is more fundamental than the universe itself, and all that we survey is a product of consciousness, rather than the other way around. Intelligence is frequently described as a channel through which consciousness acts upon the world, as if our brains were a public address system.

Also wrapped up in the bonnet of consciousness are the twin bees of emotion and free will. Feeling things (or *feeeeling* things, as one recent correspondent emphatically put it) is often seen as a uniquely human property, or at least a property of *reeeeal* life, which cannot possibly be attained by mere artificial life. The things that we *feeeel* are emotional, one way or another, so there is a logic that says artificial intelligence can never succeed, because intelligence requires emotion, and emotion is not, and never can be, felt by machines.

Even more sacrosanct than emotion is free will. At a party recently,

I found myself up against a physicist and a philosopher, who were arguing in favour of free will. The physicist's position was that free will is a consequence of the supposed disparity between the information content flowing into a black hole and that flowing out of it, and that therefore we all have singularities inside us (or at least, to be fairer to his position, that the existence of singularities was evidence for the reality of free will). As the evening wore on and the wine flowed, the argument got more and more heated, and I almost feared it would turn into a stand-up fight. Mercifully, I was rescued at the last moment by a Russian guy, who had been sitting to one side, listening intently. He asked the philosopher to do something that demonstrated his free will. The philosopher stuck his hand in the air in triumph, to which my Russian friend quietly responded: 'but you only did that because I told you to . . . '

One thing that I have learned about these topics over the years is that nobody ever changes their mind. Although I have no idea what you think about the subject I don't suppose you will change your views, and I don't suppose I will either. Whatever people want to believe, they will continue to believe. If they have to use rhetorical tricks to win their argument, some people are more than willing to do so. There is so much at stake. But ignorance is only bliss to the ignorant; everyone else has to suffer the fall-out. Getting to grips with these issues is a serious matter that should be discussed openly and honestly, because otherwise people (and non-people) are going to continue to get hurt. I know a few cows, for example, who are more than keen to establish whether they have minds of their own and thus can justifiably complain about being turned into hamburgers. I gather the vegetable community is planning on getting up a petition for much the same reason. These things matter, and we should keep looking for answers until we find them.

But hey, I have hold of the talking stick now, so let me take the opportunity of telling you a little of how it seems to me.

Let's begin by sorting out a few things about emotions. I suspect that one of the reasons people are resistant to the idea of machines having emotions is that this seems to demote something that is profoundly important to us into something trivial and mechanical. A big part of the reason for this fear is the assumption that emotions are indivisible things, and hence machines either have them or they don't. But there are five major aspects to emotion, as far as I can see, and it may help to separate these from each other.

The first aspect is the cause of most of the trouble: chemistry. One of the many design flaws of the human species, for instance, is that love and sex share much of the same hormonal chemistry. We therefore tend to confuse the one with the other, or assume that the existence of one implies the existence of the other. From this much heartache arises. One of the chemicals involved in (quite literally) producing this heartache is called oxytocin. Oxytocin is a hormone and like all hormones has a widespread and profound influence on our behaviour. It is responsible for initiating nest-building behaviour, and mediates both mother–child bonding and pair bonding, at least in other mammals. In humans it certainly stimulates milk production in lactating mothers and causes uterine contraction during labour. It is also released during orgasm in both sexes.

When we feel warmth in our heart at the sight of our children, it may well be oxytocin that generates the sensation. When we look at our partner's naked body, the warmth is usually centred rather lower down, covering both our hearts and our loins, and here perhaps oxytocin is acting in concert with something else, maybe adrenaline, to cause the dual sensation. A feeling of warmth in the loins alone can arise from less committed sexual feelings or from other exciting situations, such as bungee jumping or joy-riding, and hence there are sexual overtones or undertones to most kinds of excitement, whether we would like to admit it or not. Our complex and deeply cherished emotions are thus a consequence of simple chemistry.

This is really not the kind of news we like to be told. We resist it because it seems to demean the thing that we find most precious (love, that is, not bungee jumping), by turning it into a mere chemical reaction. But fear not. This conclusion is misleading because it is only part of the truth. When we feel love, we are indeed most probably responding to the increased production of oxytocin. The more love we feel, the more oxytocin we release into our bloodstream, and vice versa. This is relatively simple chemistry and an entirely necessary part of emotion – after all, something somewhere has to physically represent the changing levels of the emotions we are feeling. Yet there is so much more to include. What causes the oxytocin to be produced in the first place? How does the mere presence of a hormone drive us to buy flowers or write love poems? How do we feel about how we feel, when we know we are in love?

These are where all the beauty and subtlety of emotions come from. The emotion itself may be measured and represented by a simple hormone and felt in terms of this hormone's effects on our body (blushing, arousal, warmth and so on). But actually *falling* in love is a deeply complex and sophisticated cognitive process. Imagine how much computing power it takes to recognise that someone is right for us, or that they might reciprocate these feelings. The social cues and interactions that are involved in triggering the feeling of love are not at all trivial. Nor are our responses to that feeling: the way we behave may be extremely subtle and beautiful in its own right. Finally, the way that we interpret these feelings – our cognitive appraisal of ourselves – is no less meaningful, just because it stems from a chemical balance. Love feels really important and special to us and therefore it *is* really important and special. It is our high intelligence and mental subtlety that generates the value from what is, at heart, simply chemistry.

I don't know whether simple creatures such as hedgehogs fall in love, but consider fear as another example. Almost certainly, all mammals feel fear. Almost certainly the measure of that fear is provided by the same chemistry in every species. But the triggers, the behavioural responses and the interpretations of fear vary widely across our mammalian brethren. A hedgehog doesn't have many cortical maps and therefore can't detect anything beyond the most blatant reasons to be fearful. Perhaps this is why so many of them get run over. We, on the other hand, can become scared by our own wildest imaginings, of what might be lurking in the shadows, or of how we might disgrace ourselves when we step out on to the stage and everyone looks expectantly at us. A hedgehog's behavioural responses to fear are pretty limited: they can walk away or roll up into a ball. But we can give ourselves pep talks, persuade bank robbers to put down their guns or even overcome our fears completely when rationality dictates a better course of action. Who knows what a hedgehog thinks about being afraid? Horror, certainly, but perhaps nothing quite as subtle as our own interpretations of the sensation. Fear matters equally to hedgehogs and humans: it is an issue of life and death to us both; but the ways in which we think about it vary greatly in their refinement.

Future intelligent robots *must* have emotions. They must have the equivalent of chemical measures of emotional intensity, along with their concomitant bodily effects. They must have some cognitive means of detecting when to have these emotions, and they must develop a repertoire of behavioural responses that make sense. Perhaps they also need the facility to interpret these feelings and thus make offline judgements about their experiences, or have the ability to explain to us how they feel. Without these four aspects of emotion, robots cannot learn or interact socially. And there is no reason at all why they should not be given such things: simulating chemistry and implementing perception and action are perfectly achievable aims, even if we don't know how to create the more complex levels of cognition yet. For the moment, the existence of machines with cognitively weaker equivalents of our own emotions should not be seen as demoting the human spirit, any more than the realisation that we share our emotional chemistry with hedgehogs. Whether man-made systems will ever attain the subtlety of our own emotional triggers and responses is yet to be discovered, but if they do, this brings the machines up to our level, not us down to theirs.

What of the fifth element of emotion? Robots could be given the ability to perceive the necessary environmental triggers, adjust their physiology in response to them, generate a suitable reaction and rationalise their interpretation, but would they really *feeeel* anything? Would they be consciously aware of their emotional state? To be conscious is to feel; to feel is to be conscious. Even when we are only dispassionately aware, we feel dispassionate, whatever that means. Asking whether pain would really hurt them, or fear really scare them, or grief really destroy them, doesn't seem like a way through to the heart of this question, because hurting, acting scared and being destroyed are simply responses to emotion and we know that robots could do that.

I don't know the answer to this question. I simply don't know what consciousness means, and no amount of linguistic reasoning gets to the meat of the problem of how I became a 'me', and not an 'it'. But one thing I'm pretty sure about is that we need to look in the right place, and our intuitive fear of the answer often leads us to look in the wrong place – indeed at the wrong end of the scale entirely.

To believe that consciousness is some kind of fundamental principle; some underlying, indivisible, meta-something, is to resort to a belief in magic and, coincidentally, misses the point regarding the universe's most remarkable and beautiful truth. What, for example, do such vitalists (and it is vitalism, no matter how much it gets dressed up in sophisticated clothes) think the substance of consciousness looks like?

I mean, is it simple, in the sense of a spaceless, massless, timeless medium, as Descartes would have it? Or is it complex, with some structure and extension in space (even if that space is not our own physical space)? If it is simple and uniform, then how do they explain why it has complex properties? How can consciousness vary from moment to moment, person to person, consequence to consequence, if it is a uniform medium that, by definition, doesn't vary? If consciousness is not so much a metaphysical substance as an energy, or a life-force, as some people like to put it, then the same question applies. If the term 'energy' has any real descriptive value at all, it certainly can't describe what I mean by consciousness, any more than 'kinetic energy' is an explanation for why motorway pile-ups happen.

On the other hand, if this conscious meta-substance is complex – if it is convoluted and structured – then it is fundamentally a machine. That is to say, it must be the arrangement and interactions of its parts (or its convolutions) that give rise to its properties, rather than some kind of pure, formless magic. And if it is a machine, why do we need to invoke something beyond ordinary mechanics as an explanation? It is adding an unwarranted extra step to the problem, and Occam's Razor would have us remove it from the equation unless we can find some reason why this machine should have properties that no physical machine can have. Such reasons seem to be lacking, not least because we still have no idea what the limits of ordinary physical machines might be.

The mistake (and I think it is a serious and blinkered one) is to assume that all good things must come from some kind of substance. In reality all good things, and for that matter all bad things, come from the *arrangement* of substance. A phenomenon's intrinsic properties are a consequence of the properties of its parts *and how those parts are juxtaposed*. Indeed, it is the arrangement that above all defines the phenomenon – a radio receiver and a radio transmitter are made from essentially the same components; it is only their arrangement that differs. This is true for everything, from the convolutions in the electromagnetic field that we call photons, to you as a conscious being. You are not your body; you are not your brain, but you arise out of your body and brain. They create you. Consciousness is not some vague goo that perpetually underlies the universe but a relative latecomer on the scene: something emergent, which probably didn't exist a few million or billion years ago.

As such a latecomer, I am proud to be a product of the universe's greatest achievement so far. I don't want to be raw ectoplasmic jelly or a vague, primeval force field. Such concepts are trivial and mediocre, whereas organisations are endlessly rich and subtle. There are almost infinitely more ways to fit the parts of the universe together than there are parts in the first place. Many of these arrangements were not possible once. Molecules could not exist until the universe had discovered atoms; self-replicating systems (life) could not exist before the invention of complex molecules; consciousness could not exist before there were neurons and the right kinds of underlying mechanisms. This makes me, as a slice of consciousness, into the current pinnacle of the universe's organisational achievements. Consciousness should be sought at the complex and sophisticated end of the organisational scale, not the base and simplistic one.

The late Douglas Adams (and Douglas was often late, bless him; sometimes he was 'the not turning up at all Douglas Adams') had a phrase that he liked to use: 'the interconnectedness of all things'. He, above everyone I've met, had a real sense of the universe as poetry in motion: not as a thing but as a juxtaposition of events. He saw the way that it configures itself and reconfigures itself to generate novelty and beauty, and sometimes even conspires with itself to execute poetic justice. He suspected that there could even be higher forms of organisation than ourselves at work (although not in any simplistic way). At a small artificial life conference I co-organised a few years ago, he gave a brilliant extemporised talk entitled *'Is there an artificial god?'*

It seems to me that it is in the interconnectedness of all things that we should be looking for consciousness (and possibly even for god). To reduce our intricate and marvellous existence as a work of sublime organisation into a mere sizeless, massless, timeless goo is an insult and no kind of an explanation. Being machines does not belittle us, unless you have a deeply impoverished sense of what it can mean to be a machine. What does belittle us is resorting to magic – to explanations beyond our ken. If we believe that this is the case, then I think we are letting the universe down.

One final word on the subject: if consciousness is innate and fundamental, why on earth does it need so much special equipment? We know that we cannot be conscious of the external world unless we possess physical sensors and the means to interpret them. Equally, we can't even be conscious within the confines of our own mental worlds unless we have the neural machinery that I've been searching for in this book. We need a mechanism that can represent our thoughts and memories in terms of nerve activity, reflecting the causal relationships in the outside world that provide us with a grammar to think with. Without a cortex we can't imagine, and without imagination we don't exist. This is so hard to accept, largely because the one thing that our imagination cannot ever do is imagine itself not existing. Every night, during the deep sleep between our dreams, we cease to imagine and hence cease to exist. But since we aren't there to witness these episodes, our conscious existence seems to us like an unbroken stream, even though we know formally that time has passed by without us. Finally, a crucial component of consciousness is our ability to have feelings about what we witness, whether inside or outside our heads, and in order to feel we need an emotional system. So, if consciousness is fundamental rather than emergent, it seems remarkably odd to me that it had to wait billions of years for the evolution of higher brains before it could have any consequence.

I'm merely offering these thoughts for your consideration (with the humble recognition that you've probably already thought of them for yourself). I'm not trying to ram them down your throat. If you don't like the implications (and I'm not sure I do either) then you are free not to believe a word of it. What right have I to upset you? On the other hand, some people do try to ram their sometimes ill-considered beliefs down other people's throats, just because it makes them feel better. In the previous chapter I mentioned a small group of people who killed themselves in the presumably confident belief that their consciousness would emerge into a better place. But in so doing, they took away the consciousness of a much larger number of people, many of whom believed nothing of the kind – people who really felt they needed their brains and their emotions to remain intact in order for their being, or at least their well-being, to persist. We are entitled to our own beliefs just as long as we don't inflict them on other people; the minute they become public they have to be justified. The foundations of justice and freedom lie in working out what consciousness really is, and not what we would like it to be. One of the best reasons for building robots like Lucy is to help us articulate these issues and hopefully make a little more sense of them.

Justice and freedom bring me to my last topic, which is no less stressful to think about and no less urgent either. Before I leave this chapter I suppose I have no choice but to return to the question of free will.

My own position is that there is no such thing, but that we will, can and must act *as if* there is such a thing. Even though I think the future is completely inevitable, I also know it to be utterly unpredictable, and therefore it doesn't matter two hoots to me that what is going to take place will happen, whether I like it or not. The point is, I don't know what is going to happen, and nor does anyone else, so what difference does it make? I'm just as excited to find out, no matter how inevitable it may be. Anticipating the future is, as I hope I've begun to explain, what brains really want to do.

The only thing I can say to people who believe in the reality of free will at a fundamental physical level, is this: if you make what you believe to be a free choice, what was it that caused you to decide one way and not the other? If your choice was purely random then this is a pretty dull and lamentable kind of freedom. If your choice was lawful (by which I mean it was a logical consequence of your state of mind, your background and recent events in the outside world) then you were acting inevitably, and there was no choice at all. If a free choice is one that is neither random nor consequent on something, then what on earth is it? What word can you offer to describe it? 'Free' is not enough of an answer; playing word games won't wash with me. Nor will resorting to semi-magical get-out clauses like the quantum superposition of states. Nor will recourse to many-universe theories, in which both choices happen. Nor, for that matter, will recourse to religion and hence to influences from outside our universe. It doesn't matter what you try, the question is still the same: if a choice is not acausal (and therefore totally random) and not causal (and therefore inevitable), what is it?

Until someone can answer this question, I have no choice but to disbelieve in free will as an innate physical concept, although I'm still perfectly happy to accept it as a social construct and an ethical stance, because the word 'cause' has an altogether different set of connotations in the social sphere. It all depends on your level of description, and that, I think, is the general point I'm trying to make in this chapter.

I suppose, when it comes down to it, I could simply choose to believe in free will because it makes me feel better, despite the logical absurdity of it. However, I think I'd always have that nagging feeling that this was what I was going to do all along.

EPILOGUE

Can't quote Kant: finding my own Nietzsche

Alan bloody Harwood! It was all his fault.

As I was nearing the end of the work chronicled in this book, I found myself suffering from an ever-growing sinking feeling. Its chief cause was obvious, correlating precisely with the depletion of our bank balance as my mid-life fantasy to understand some of the operating principles of the brain took its inevitable toll on our life savings. But there was also something oddly familiar and infuriating about the feeling, which I couldn't quite pin down or describe. And then a long-forgotten name rose up from the darkest recesses of my memory: Alan bloody Harwood.

We were about nine years old, I think, and our teacher had decided that the class would perform a piece of music as part of the school's Open Day. Of course, the girls (being almost grown up by that age and hence much smarter than the boys) got to play the recorder. They were clever enough to read music, move their fingers and breathe, all at the same time. What's more, they were willing to take their recorders home and practise, without feeling obliged to slope off into the woods and play war games instead. For the rest of us, though, it was a matter of casting around for something that we could bang, shake, or at worst simply avoid tripping over.

Alan and I were assigned to a pair of glockenspiels. I adore the glockenspiel: such a pure, warm sound. But could I actually play one? Not a chance. We couldn't take the instruments home, so I tried my very best, during the limited amount of classroom time allocated to rehearsal, to link up a few of the right notes into something resembling the right order. Sadly, I couldn't manage to play them at anything like the right speed. I remember the tune well: *The Skye Boat Song*. Sit me in front of a keyboard today and I can still tap out the sixteen notes that I managed to burn into my brain at the time, but after that, everything is a complete blank. Alan, on the other hand, had better things to do with his childhood than delve into the arcane mysteries of musical notation. For him the sheer tactile joy of slamming rubber hammers on to metal keys was enough reward in itself, so rather than learn any of the actual notes, he simply thrashed away with both hands, clouting the chromed keys indiscriminately, with a manic grin stretching from ear to ear.

Our teacher toured the group, peering over her pince-nez to judge how well we were doing (prowling around and peering at people was what teachers were trained for in those days). There I was, face screwed up with concentration, doing my very best but already three notes behind by the end of the second bar. She waited and watched, while the countdown of my memorised notes fell inexorably from sixteen to zero, to be followed by an endless awkward silence, like an expensive firework rocket on a very damp night. Meanwhile, there was Alan, smashing away happily at random, arms akimbo. Perhaps the woman was tone deaf or something, but it was at this point my teacher made one of those decisions that seem so inconsequential, but which, unbeknownst to them, can destroy children's souls. She put Alan (bloody) Harwood at the front of the orchestra, where our visiting parents could admire his entirely sham virtuoso performance, and she sent me right to the very back, where I played my sixteen notes to an invisible audience from behind a pillar.

For such a minor insult to my childhood dignity, this episode taught me a deeply cynical lesson: it illustrated the disheartening extent to which people will intuitively favour image over substance. The moral of this tale is relevant to my topic in two ways, one of them intellectual and one personal. Both of them relate to how we do science.

After three years of hard work, I have managed to build a robot that is just about capable of learning how to point at a banana. Eight computers, tens of thousands of lines of computer code and a network of more than a hundred thousand simulated neurons, and all it can do is point at fruit. Big deal. Were Alan Harwood to build robots, he would have the sense to replace all this unnecessary complication with a simple instruction like 'point to the yellowest half of the image'. So why have I gone to such absurd and painful trouble to do everything the hard way, and what, if anything, does it prove?

One of the ideas I've been trying to get across in this book is that intelligence is not something we are born with but something we develop. Our brains pull themselves up by their own bootstraps, and they do so from the most humble beginnings. They have to, because our highest mental concepts can only exist by building upon simpler, more tangible ideas such as friction, volume and shape. If AI can build a machine to play world-class chess (to take a rather iconic example) but the same machine is unable to pick up the pieces when they fall over, or tell a rook and a bishop apart simply by looking at them, then I think something is seriously wrong with the underlying theory. The results are impressive, as far as they go, but they are not built on the right foundations and will never lead to real, general intelligence.

This is a problem for the public (and sometimes even for funding agencies) because it can be so difficult to distinguish between what is real and what merely looks real. People are apt to find themselves unrealistically impressed, or sometimes unnecessarily worried, by 'progress' in AI that in reality is far more superficial and limited than it seems. I don't for one moment mean to imply that scientists are intentionally misleading the public. I'm perfectly aware of the solid rationale and excellent intentions that lie behind the vast bulk of research projects. I'm simply pointing out that initially impressive behaviour is not necessarily a good guide to ultimate fruitfulness, and sometimes it is hard to tell the two apart.

Unfortunately, I am sometimes accused of faking it myself. A respected scientist recently dismissed my project with an airy wave of the hand, saying something to the effect that any idiot can build a robotic puppet that simply *looks* like it is intelligent. Well exactly! That is the point I'm trying to make. But my aim with Lucy is certainly not to focus on image rather than substance – if it were then I chose a pretty stupid way of programming her to point at bananas. I think this commentator might in part have been misled by my whimsical tendency to decorate my robot with silly bits of fur and suchlike, and that's something I'll just have to live with. I think it's important to engage the public in what scientists do, and I know that an attractive, anthropomorphic machine is more likely to catch the eye and lead people to ask deep questions than is the average lab robot.

A second reason I come in for criticism is for making outrageous claims. I swear I do nothing of the kind, but people will insist on reading between lines that don't exist. According to some people I am trying to create human-level intelligence; according to others I believe that Lucy will become conscious. It's not my fault if these people are unable to distinguish between human-*level* intelligence and mammal-*like* intelligence (in other words, the kind of general learning capacity that cortex-enabled mammals have and insects don't). Nor is it my fault if journalists (with the kindest of intentions) over-enthusiastically turn my attempts to understand the biological origins of conscious thought into a bald claim that I'm trying to build a conscious machine. I wish people would listen to what I say, rather than what they think I say.

Finally, there is the presumption that I am arrogant enough to believe I can solve all of the problems in one go. Instead of focusing on one small, tractable aspect of the problem in the traditional manner, I've had the nerve to build a complete robot, with muscles, hearing, voice, vision and everything else, as if I expect to solve all of the problems of cognitive science by myself. But this is a mistake. I'm not doing it for show, or out of bravado. I have given Lucy a wide range of senses and motor systems for one specific reason: to help me recognise underlying principles that are independent of any particular sense. Vision, hearing and muscle co-ordination (not to mention the more abstract processes of thought) seem like very different problems on the surface, and are usually studied by entirely different groups of people, but nature solves them all using a single machine. Cortex is cortex is cortex. My goal is to understand how cortex develops its ability to specialise, and for that I need to look at the problem from as many angles as possible, or I shall find myself blinkered by the particulars.

Which brings me to the second point I want to make. I don't believe intelligence makes any sense in the absence of a *complete* organism, embedded in a rich and demanding environment. To some extent, therefore, I think we really *do* have to solve many of the problems at once, since none of the individual parts are comprehensible when considered in isolation. Who would be able to understand the logic behind a spark plug if they had never seen a piston, and how could they see the point of a piston unless it was attached to a crankshaft? The Lucy project was therefore inevitably going to be something of a magnum opus in research terms: a slow to mature, mega-project that needs to be tackled in a very holistic, single-minded, non-disciplinary way, rather than an incremental, easily measurable, collaborative one. Unfortunately, this is not what policy makers, institutions or indeed our education system seem to want to encourage.

Incidentally, I use the phrase 'non-disciplinary' rather than 'inter-disciplinary' rather pointedly. Someone who was writing an article on Leonardo da Vinci recently asked me whether I thought a polymath like Leonardo would be able to flourish if he were alive today. My answer began with the observation that I don't think Leonardo was a polymath. I think he was interested in one thing, and one thing alone: how does nature work? He therefore used whatever means seemed appropriate in order to tackle this one question. The fact that we today regard painting, sculpture, philosophy, experimental science and invention as radically distinct disciplines is a much later accident of history, and a rather unfortunate one at that. 'Non-disciplinary' denotes a healthy disregard for all such artificial and constrictive boundaries.

Yet there is little encouragement (in this country at least) for such a Third Culture attitude. The term 'Third Culture' was coined by John Brockman as a response to C. P. Snow's 'Two Cultures' of art and science. Unfortunately, the division into two cultures doesn't even *begin* to describe the over-compartmentalisation of human knowledge and intellectual endeavour that exists, and the enormous pressure that culture in general puts on us all to specialise.

This is not a good thing. We do need specialists, of course. Many things would not be possible if there were no people able and willing to know a huge amount about relatively little. I wouldn't want to live in a world where people with only a vague understanding of radioactive decay design nuclear power stations, and I wouldn't fly behind a pilot who didn't have a pretty detailed understanding of aerodynamics. Nevertheless, we also need generalists, and our education system, our funding mechanisms and our natural tendency to carve the world up into neat categories at the drop of a hat, all militate against such non-disciplinary thinkers.

Innovation comes from seeing things differently. A simple shift of viewpoint can work wonders, and one excellent way to achieve such purity of mind is not to educate it, or rather not to fill it with too much conventional wisdom arranged in conventional ways. Education (as most educators know perfectly well) should help people learn how to think for themselves, but in practice the only efficient way to jump the hurdles of tests and exams is to fill their heads with the thoughts of others. Since this is the diary of one person's uneducated, unspecialised attempt to understand the way we (and robots) learn, I would therefore like to take this opportunity to put in a good word for the autodidacts and independent thinkers of this world.

For the past few years, about once or twice a week on average, I've had emails from complete strangers who proceed to engage me in deep, stimulating discussion or argument about something – more often than not about consciousness. After three or four exchanges of views, the writer will gain enough confidence to say something like 'Oh, by the way, I'm only educated up to the Sixth Grade', or 'Of course, I left school at sixteen when I was told I was backward', or 'I worked my ideas out while I was pushing my rubbish cart around'. What a waste of talent! Opportunities to change the way we think about the world are too often handed out in accordance with how many paper qualifications people have, despite the fact that these qualifications simply prove how good their owners were at absorbing and regurgitating the old ways of thinking. Meanwhile, anyone who ploughs their own furrow is left to pull their own plough.

I didn't really intend this book to be so autobiographical or so first person. There's a risk that if you don't like my personality, you won't listen to my ideas. But science *is* personal, especially when you practise it on your own, using your own hard-won and rapidly dwindling resources. Scientific endeavour is so often portrayed as being completely dispassionate, or if it is given a personal gloss then this is to suggest that the process is fiercely competitive, one lab vying with another for precedence and hence superiority in the fight for fame or funding. But this portrayal misses the point. It is the *ideas* that really kindle the passions. Mankind evolved as a species of scavenger, discovering an unexpected source of nourishment here, a novel means of shelter there, a bush back in the corner where they could steal a little sex without the alpha male knowing. Curiosity is therefore built into us, for only the curious survive in a scavenger's market. Knowing new things, applying old knowledge to new problems, seeing the common features underlying a confusing and disparate world, are what we are passionately driven to do. It doesn't matter whether it is noble; it is within us. Unfortunately, satisfying one's curiosity beyond the immediate needs of friends and family is a luxury.

Looking back over these pages and the past three years of my journey, I can see how my mood began to rock ever more violently from side to side, as my depth of understanding and intellectual confidence in the project advanced on the one hand, while our bank balance simultaneously retreated on the other. I felt like a passenger on the *Titanic*, edging steadily but with frustrating slowness towards the security and splendour of New York, while at the same time becoming frighteningly distant from the safety of home. Perhaps the high seas are no place for a self-educated, amateur scientist after all. As I was writing the last few chapters the final crunch seemed only moments away, and I found myself staring helplessly at an approaching wall of ice. At any moment I expected the band to strike up their final rendition of the *Skye Boat Song*, on recorder and glockenspiel, as my family and I started to sink silently and unnoticed into the frozen waters of bankruptcy.

But I have some good news to report. Just in the nick of time a lifeboat arrived, in the welcome shape of NESTA – the UK's National Endowment for Science, Technology and the Arts. NESTA was set up specifically to encourage and help people like me; people who don't fit the conventional funding categories. They know perfectly well that what I'm doing is high-risk and a little bit flaky, but they also have faith that something useful will come out of it, even if from a completely unexpected angle. It always has before, yet this level of support is a rare experience in my life.

I would therefore like to record my heartfelt thanks to everyone at NESTA, particularly Venu Dhupa and Cath Wilkins, for having faith in me and for filling a much-needed gap in the funding system. While I'm thanking people, this might also be a good place to thank the people involved in this book, since that also directly helps to fund my research. Specifically, my thanks go to my editor, Richard Milner, and my copy editor, Ann Grand (who is also my wife and thus deserves double thanks for allowing me to indulge my curiosity instead of getting a proper job). Finally, thank you, the reader, for buying it.

For the first time since I started this project, someone else is picking up the tab (for a year, anyway). So, for the past few months, while the rest of this book has been going through the editing process, I've been busily designing and experimenting with a completely new body for Lucy (see plate 8). Lucy Mk I was a sad little hotchpotch of scavenged parts and cheap-but-dirty solutions to problems, and it is very difficult for a girl to be intelligent when she can't see properly, hear properly or have much effect on the world. But now I can afford decent tools and components, such as a lathe, a rapid prototyping system and better development software. It means going right back to square one for a while, but a new body will be a huge improvement and greatly increase my ability to test out my ideas.

Lucy Mk II is going to be bigger, shinier and more robust. She will have far more realistic muscles (although the limitations of electric motors are still driving me crazy) and at last she will have legs. I don't know yet whether these legs will be strong enough to enable her to learn to crawl, but if all else fails, I can give her a small electric tricycle to ride around on, inside which I can store the greatly enhanced (and consequently bulky) computers that I can now afford. These more powerful computers will enable her eyes to resolve more detail and see in colour. She will also have two eyes instead of one, so I can play with ideas about how the brain deals with depth perception. Her hearing too will be enhanced, as will her voice. At last she will be able to say more than 'arp', and I can begin to explore ways in which her brain might construct useful relationships between the muscles she contracts in her vocal tract and what she actually hears herself say.

Above all, the enhanced computer power will enable me to simulate many more neurons simultaneously. I can finally integrate all of the elements I had to develop separately during the course of this book – the control of visual saccades, the spatially invariant visual recognition, the servo-driven sequencing of muscle groups – and from there move on to greater things. I'm especially looking forward to implementing the ideas I outlined in chapter 17 about self-organising morphs between coordinate spaces, although this time using proper functioning cortical maps, rather than a simplified model.

Whether the result ends up any closer to being an intelligent, autonomous entity after all this effort remains to be seen, but at least, for a while, the project is back on track. If I come up with anything interesting, I'll let you know.

I really wasn't expecting a happy ending to this journal of events, but as an amateur biologist I should have known better: once conception has taken place, every woman knows that she can't stop the baby from growing inside her, and so she learns to prepare herself for the inevitable. One of the signature differences between the sexes (to make a sweeping generalisation) is that women intuitively know this and hence learn to practise self-control, while men expect to remain in control of the world itself, rather than of their responses to it. In a strange and rather disturbing way I've begun to feel like I'm pregnant myself, as if I were heavy with robot, so maybe I should simply learn to live with the inevitability of it all. Once conceived, life is unstoppable. It is by far the greatest self-organising force in the universe. Things usually work out fine for it in the end.

Being male, I had always assumed that I was the one doing all the work, trying hard to bring Lucy into the world. But I'm beginning to wonder if I'm the master or just a willing slave. Perhaps this whole adventure was really part of Lucy's plan all along. It wouldn't surprise me. Girls can be so manipulative!

INDEX